To Paul

from

Thank you for many inspiring discussions during my stay & also thank you also for your warm hospitality

Abe Ome. 1986

IEE ELECTROMAGNETIC WAVES SERIES 22

Series Editors: Professors P.J.B. Clarricoats
E.D.R. Shearman and J.R. Wait

Target adaptive matched illumination RADAR:

Principles & applications

Previous volumes in this series

Volume 1	**Geometrical theory of diffraction for electromagnetic waves** Graeme L. James
Volume 2	**Electromagnetic waves and curved structures** Leonard Lewin, David C. Chang and Edward F. Kuester
Volume 3	**Microwave homodyne systems** Ray J. King
Volume 4	**Radio direction-finding** P. J. D. Gething
Volume 5	**ELF communications antennas** Michael L. Burrows
Volume 6	**Waveguide tapers, transitions and couplers** F. Sporleder and H. G. Unger
Volume 7	**Reflector antenna analysis and design** P. J. Wood
Volume 8	**Effects of the troposphere on radio communications** Martin P. M. Hall
Volume 9	**Schumann resonances in the earth-ionosphere cavity** P. V. Bliokh, A. P. Nikolaenko and Y. F. Filippov
Volume 10	**Aperture antennas and diffraction theory** E. V. Jull
Volume 11	**Adaptive array principles** J. E. Hudson
Volume 12	**Microstrip antenna theory and design** J. R. James, P. S. Hall and C. Wood
Volume 13	**Energy in electromagnetism** H. G. Booker
Volume 14	**Leaky feeders and subsurface radio communications** P. Delogne
Volume 15	**The handbook of antenna design, volume 1** Editors: A. W. Rudge, K. Milne, A. D. Olver, P. Knight
Volume 16	**The handbook of antenna design, volume 2** Editors: A. W. Rudge, K. Milne, A. D. Olver, P. Knight
Volume 17	**Surveillance radar performance prediction** P. Rohan
Volume 18	**Corrugated horns for microwave antennas** P. J. B. Clarricoats and A. D. Olver
Volume 19	**Microwave antenna theory and design** Editor: S. Silver
Volume 20	**Advances in radar techniques** Editor: J. Clarke
Volume 21	**Waveguide handbook** N. Marcuvitz

Dag T. Gjessing

Target adaptive matched illumination RADAR :

Principles & applications

Peter Peregrinus Ltd. on behalf of the Institution of Electrical Engineers

Published by: Peter Peregrinus Ltd., London, United Kingdom
© 1986: Peter Peregrinus Ltd.

All rights reserved. No part of this publication may be reproduced, stored in a retrieval system or transmitted in any form or by any means—electronic, mechanical, photocopying, recording or otherwise—without the prior written permission of the publisher.

While the author and the publishers believe that the information and guidance given in this work is correct, all parties must rely upon their own skill and judgment when making use of it. Neither the author nor the publishers assume any liability to anyone for any loss or damage caused by any error or omission in the work, whether such error or omission is the result of negligence or any other cause. Any and all such liability is disclaimed.

Gjessing, Dag T.
   Target adaptive matched illumination radar:
principles and applications.—(IEE
electromagnetic waves series; 22)
1. Radar
I. Title   II. Series
621.3843   TK6575

ISBN 0-86341-057-X

Printed in England by Short Run Press Ltd., Exeter

Contents

		Page
Preface		*vii*
Acknowledgments		x
1	**Introduction**	1
2	**Matched illumination: basic principles**	8
	2.1 Detection/identification of objects by wave-number matching (spatial resolution)	9
	2.2 Transverse distribution of scattering elements: spatial correlation of scattered waves, synthetic aperture method	24
	2.3 Characterization of scattering centres by polarimetric methods	36
	2.4 Temporal properties of radio waves scattered from a flexible body in motion	44
	2.5 Space/time coherence of electromagnetic waves scattered back from an object: measurement of object rigidity	51
	2.6 Influence of absorbing surfaces: molecular resonance phenomena and illumination matched to surface material composition	55
3	**Characterization of scattering objects by means of an adaptive multifrequency radar; experimental examples**	68
	3.1 A brief description of an experimental adaptive radar system	73
	3 2 Illustrative examples from experiments on rigid objects	74

4 Classification of vegetation based on size, shape and motion pattern — 84
    4.1   Characterization of vegetation on the basis of shape and size — 85
    4.2   Characterization of vegetation on the basis of motion pattern — 89

5 Sea clutter background: characterization of ocean surface — 92
    5.1   Motion pattern of the sea surface irregularity structure — 94
    5.2   Ocean wave spectra, wave height and wave length — 98
    5.3   Interaction mechanisms between plane electromagnetic waves and sea surface irregularities: basic principles — 100
        5.3.1   Multifrequency radar principle and frequency difference matching — 110
        5.3.2   Some illustrative experimental examples — 116

6 Ship targets against a sea clutter background — 131
    6.1   Spatial signature (wave-number matching) — 131
    6.2   Temporal (Doppler) signatures — 133
    6.3   Space/time (mutual) coherence: target detection enhancement based on the measurement of rigidity — 138

7 Detection of ship wakes by matched illumination — 144
    7.1   Directional properties of gravity waves generated by ships: detection of ships by matching the illumination to the ship's wake — 146
    7.2   Directional properties of internal waves generated by a disturbance of a surface ship — 149
    7.3   Coupling mechanisms between internal waves generated by a ship and surface irregularities — 153

References — 159

Bibliography — 163

Preface

Few generations of radio scientists have had opportunities such as those we are witnessing. Objects or phenomena with properties that were hitherto unidentifiable, even in the laboratory, can now be studied using electromagnetic waves from remote platforms. The radio scientist of today owes this privilege to previous generations, who laid the foundations of technology and methodology on which can be built new and complex solutions not hitherto conceivable.

The achievements in radio science over the last decade have been promoted by several factors. At first, progress was stimulated primarily by interest in communications and by military needs. A period followed in which space research was the driving force. At present the main interest is in problems concerned with the environmental sciences, the earth's resources, pollution and conservation.

In the field of information engineering, very refined signal processing schemes based on classical concepts can today be realized by a powerful set of flexible computer elements and a library of software for a wide variety of fundamental problems.

Likewise, new technology emerging from innovations in solid-state physics has given us a large selection of components and subsystems in all frequency bands from which dedicated diagnostic tools can be built. For example, the preceding generation of scientists had little more in the line of microwave illumination than the magnetrons, whose frequencies were largely fixed and determined by their mechanical geometry. Today we are blessed with solid-state microwave sources whose amplitude and phase (frequency) can be controlled by microprocessors in a predetermined manner, thus providing us with illuminators which can be matched (adapted) to the object of interest.

Interaction mechanisms between acoustic waves in piezo-electric materials and electromgnetic waves (SAW devices) will soon enable us to implement high speed analogue processing concepts.

In addition to being faced with challenging advances in basic science and engineering, we are living in a decade where our achievements in science and engineering can be transformed into relevant and stimulating applications. At no time in the history of science has the time span between basic research and mission oriented applications been as short as today.

We have every reason to be optimistic, every reason to adopt a progressive attitude, in approaching the task of developing an optimum system for detection and recognition of general objects against a background.

The basic philosophy adopted – well known from other fields of science – is as follows. Most, if not all, of the objects that our radar will face in our terrestrial environment are known in advance. We possess all the appropriate information about shape and size, rigidity, surface material composition and motion pattern. In essence we require a very limited amount of information: is the object of interest on the scene or is it not, and what is the position of the object?

Few existing radar systems are flexible enough to make use of all this *a priori* information about the target. We know from classical information theory that a communication system requires zero bandwidth to transfer a known message. By the same token, it takes zero bandwidth for a radar to detect a known object. Obviously, the more *a priori* information about the object we possess, the smaller can be our bandwidth, or, for a given bandwidth, the quicker we can detect and identify the object.

The target adaptive matched illumination radar makes use of these classical concepts from information theory. It is believed that although the concepts are classical, the implementation hitherto has been hindered by technological shortcomings. These are now, as we shall see, overcome.

In applying radio science principles to the problem of the detection and classification of objects or phenomena with hitherto unidentifiable properties, the need arises for a unified treatment of these principles and their applications. This book endeavours to satisfy this demand. It attempts to provide a brief and coherent outline of the underlying theory and of the basic principles behind present-day adaptive radar matched illumination methodology, and to apply the same fundamental concepts to the problems of the target and of the environment against which the target is viewed. From such a unified treatment it is possible to synthesize the optimum detection system – illumination (matched illumination) and 'matched filter' reception – taking into consideration the characteristics of target and background. What this means in essence is that if we know the signatures of the target of interest and of the background, we can put these signatures into the memory of a computer and instruct the computer to adapt the radar illumination as well as the processing of the received signal to the target in an optimum manner.

The book is intended for advanced students or scientists with good all-round knowledge in the field of electromagnetics and inversion theory, but it does not assume specialist knowledge. An effort has been made to use simple first-principle first-order mathematics so as to produce a book that can be read from cover to cover without supporting literature. It is the hope of the author that the book, if read in this way, will reveal some of the main ideas of target adaptive matched illumination radar as applied to the challenging problems of today. Furthermore, it is hoped that the reader, like the author, will find stimulating the current developments within the general field of radio science.

Dag T. Gjessing

Kjeller, Norway

Acknowledgments

Looking back upon the work in preparing this book, I feel a deep sense of gratitude to a number of people: colleagues, assisting staff and employers. To no one, however, do I owe a more sincere thanks than to my secretary, Eva Roedsrud, and to my associates Svein-Erik Hamran, Jens Hjelmstad and Eldar Aarholt.

Only through their expert assistance, their conscientious co-operation and their loyal support was it possible to complete this book during the short time available. Questions helped to remove obscurities, discussions gave stimulus and added momentum, and the companionship contributed to satisfaction and animation.

Among colleagues at my earlier affiliation – the Norwegian Defence Research Establishment – I am particularly indebted to Karl Holberg. His wisdom and his creative and responsive mind have been a continual source of inspiration and crucial encouragement.

Likewise, I wish especially to thank my colleagues at the Royal Norwegian Council for Scientific and Industrial Reserch, Hans C. Christensen, Jan M. Doederlein and Andreas Tonning, and my colleagues at Tromsoe University, Ove Bratteng and Kristian Dysthe, for stimulation and animation and for loyal support at crucial moments.

I also wish to express my appreciation to Ruth Sand who tolerantly and very ably handled the arduous task of typing most of the manuscript, and likewise to Marit Ekstrand and Aud Nymoen, who prepared all the artwork.

Finally, I wish to thank my family, my wife Toril, daughter Randi and son Trygve. The effort involved in preparing this book, during evening hours, has also required their loyal support.

Some of the material and illustrations I have used in earlier publications. The permission by the various journals to reproduce the material here is gratefully acknowledged.

I also acknowledge with thanks the encouragement, by Professor J. R. Wait, to write this book.

Chapter 1

Introduction

Since the early days of radio science, electromagnetic waves have been used as a diagnostic tool in a wide spectrum of applications. It is only in recent years that the term 'remote sensing' has been introduced as a loosely defined common denominator, primarily in connection with passive multicolour photogrametry from satellites. Now a large congregation, mainly from the user community, ride on the 'remote sensing bandwagon'.

This book aims to give a description of general wave interaction phenomena based on fundamental laws of physics. To ensure a general physical understanding, let us at this early stage introduce some of the more important concepts on which the later chapters of the book will be based.

With this objective, it may be appropriate to present an artist's conception of the radio signature domains that we have at our disposal. These domains are illustrated in Fig. 1.1.

Note first one main distinction, namely that of passive and active methods. In the passive mode we are confined at the observation point to measuring the angular distribution of power only. Whether this angular power distribution is a result of thermal radiation from the object in question, governed by Planck's law of radiation, or by scattering of incoherent waves originating from the sun; we are deprived of information about phase angle and polarization. Hence, all we will have is information about the absorption/emission coefficient and temperature in a plane normal to the direction between the object and observation point. Lacking phase information, we cannot say anything about the distribution in depth (range) of the scatterng body, nor can we say anything about motion pattern or polarization characteristics of the scattering/radiating body. The passive mode (measurement of radiation temperature) is depicted in the lower left-hand side corner of Fig. 1.1.

Fig. 1.1 depicts in a rather self-explanatory manner the various signature domains that we have at our disposal. The target, in the form of a ship, can be detected, and, under certain conditions, also to some extent characterized on the basis of its footprint (Kelvin wake, stern waves and turbulent wakes, generation of internal waves, upwelling of cold water, stimulation of biological

2 Introduction

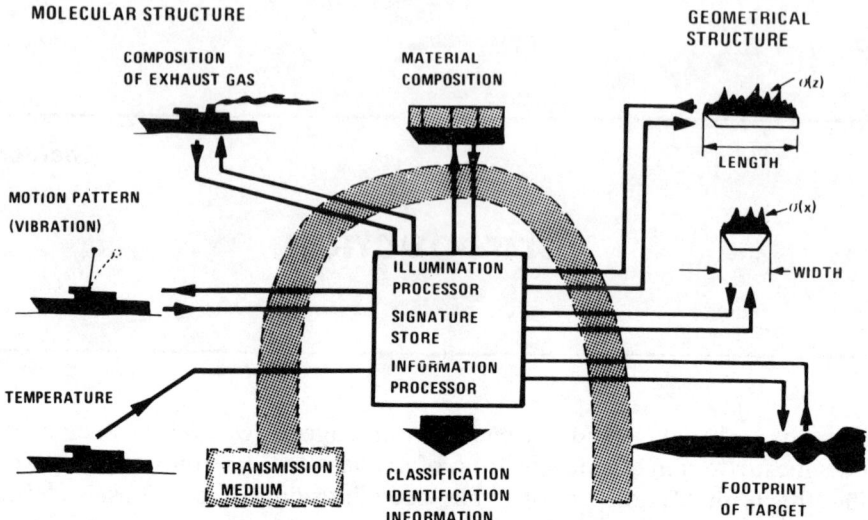

Fig. 1.1 An object can be characterized in several signature domains. Based on the footprint a target can be detected and often also characterized. Based on Maxwell's equations and direct and inverse methods, the shape and size can be determined. Using a tunable laser, molecular resonance phenomena can be revealed and material composition determined. Based on Doppler measurements, the speed, scale selective vibration pattern and rigidity through space/time coherency can be obtained

phenomena such as plankton etc.). Furthermore, the figure indicates that the target can be characterized by means of the geometrical structures revealed by direct or inverse scattering methods on the basis of Maxwell's equations; based on surface spectroscopy using appropriate tunable lasers the material composition and composition of exhaust gas can be revealed, and by measuring relative phase as a function of time the motion pattern can be determined.

In an earlier book by the current author (Gjessing, 1978a) a survey of the various detection/identification criteria was given. In this contribution we shall confine ourselves to methods based on scattering of electromagnetic waves using coherent polarimetric techniques for the purpose of measuring:

1 Shape and size of scattering surface (liquid or solid).
2 Scale selective motion patterns (speed, vibration, dispersive and non-dispersive phenomena).
3 Scale selective space/time coherency and thus rigidity.
4 Polarization properties of the scattering centres constituting the scattering body.
5 Molecular structure.

In a passive radiometric mode, the radiation temperature distribution can be determined if the emissivity is known.

As emphasized in Fig. 1.1, whether active or passive methods are employed, the intervening propagation medium may play a dominating role through frequency selective absorption mechanisms, through turbulence giving rise to phase perturbations causing blurring, and through motion phenomena resulting in Doppler distortions.

Hence, when designing a system for characterization of scattering objects, not only the properties of the object itself and the background against which the object is viewed have to be considered, we must also consider the filter constituted by the transmission medium. This question is addressed in some detail in a recent book by the author (Gjessing, 1981b).

In summarizing the introduction of target signatures in relation to background giving additive noise and in relation to the propagation medium giving multiplicative noise, Fig. 1.2 is presented.

We have now reached the stage in this general introduction where the term 'target adaptive matched illumination radar' can be introduced. This book is dedicated to the following concept:

In most practical cases one has ample *a priori* information about the target of interest, based on which an optimum 'matched illumination – matched reception' system can be designed. Knowing something about shape and size and about motion pattern and rigidity of an object in relation to the background against which the target is viewed, an optimum illumination and detection system which adapts itself to this target can be designed.

The more orthogonal independent dimensions we have at our disposal, the less detailed does the description in each signature domain have to be. Such a multisensor data-fusion system makes it possible to recognize an object with

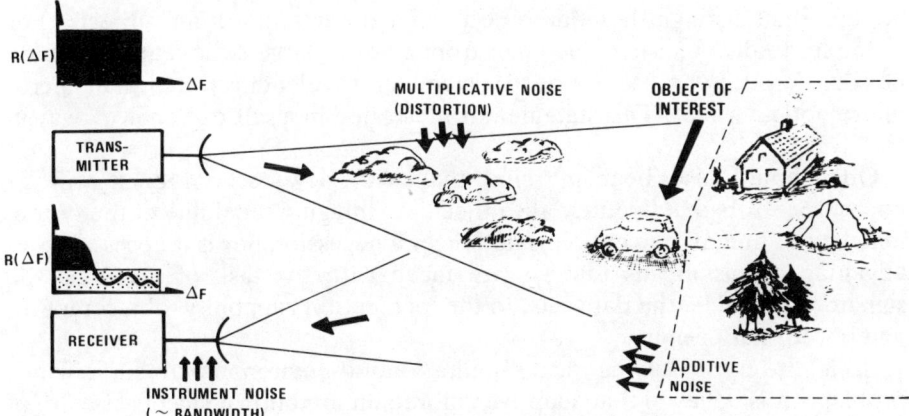

Fig. 1.2 *To optimize the contribution to the received signal from the target (signal/noise ratio) the signature of the target must be considered, and in addition, that of the background giving additive noise and of the intervening propagation medium giving multiplicative noise*

high certainty (low error rate) by considering only 'the most significant bits' in each signature (sensor) domain. Hence, the bandwidth (spatial resolution) in each of the orthogonal domains need not be extreme.

What this in essence means is that we can tell the illumination computer everything that we know with certainty about the target of interest, thus confining the task of the radar to collecting the *difference* between large amounts of information.

When studying, as an example, ships on the sea surface or the sea surface itself, an image of the scene does not represent a desirable solution. If one is interested in ships, information about position x,y about velocity v, and about shape and size is desirable. If the interest is confined to the sea surface, then one would like to be given a statistical description of the sea (directional wave spectrum) rather than an image.

Present-day technology and concepts make it possible to design an *automatic* (adaptive) detection/recognition system which can be carried without an operator on a small platform relaying the required processed information directly to the user (coastguard for maritime surveillance, oil drilling platforms for wave assessment and forecasting).

Although one may well need decimetric resolving power for the purpose of identifying a target, it is most unlikely that such an accuracy is needed when determining the position of the target. A conventional non-adaptive SAR system, for example, resolves the whole scene with the same resolution as that required for the target. Obviously this does not represent a desirable situation.

What we seek is an adaptive system that extracts *adequate* information for a given task rather than the ultimate information. In order to obtain adequate information the object of interest will have to be illuminated by electromagnetic waves having wavelengths (bandwidth) which allow it to resolve the feature that distinguishes our object from other appropriate objects. For example, to distinguish an elephant from a horse, large-scale features should be considered. For these one needs decimeter wavelengths rather than micrometre optical waves. This statement is illustrated in a self-explanatory way in Fig. 1.3.

One should also bear in mind that with new technology it will be microprocessors that 'identify' the object. An imaging capability of the system matched to human perception is no longer a necessity, nor is it necessarily an advantage. This means that we are faced with the task of matching the signature domains, the data sets, to the 'perception capability' of a computer and not to that of man.

In addition to solving acute military and environmental surveillance problems it is believed that adaptive tailored illumination concepts also are of importance for robot vision in connection with quality control and object classification.

Hence we are faced with the consideration of three filter functions. One is constituted by the transmission medium between the observation platform and

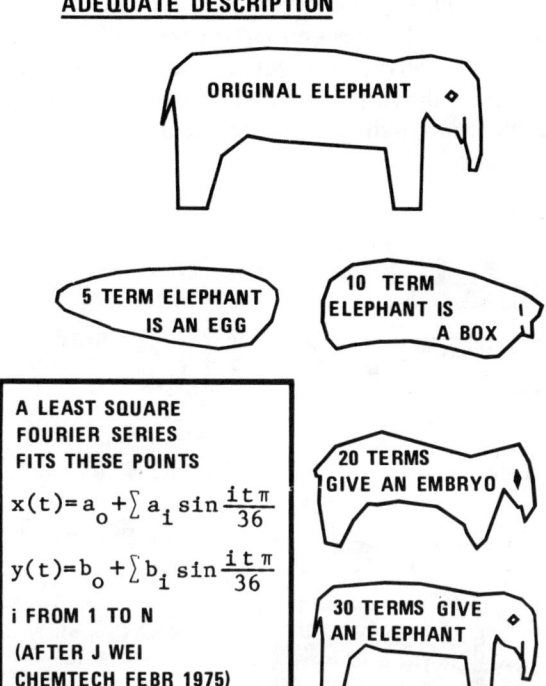

Fig. 1.3 *In most practical cases the challenge is to provide adequate description of an object so as to be able to recognize it, i.e. to distinguish between several possibilities. We should therefore illuminate the object in an optimum manner and with adequate bandwidth (wavelength)*

the target (this intervening medium gives rise to multiplicative noise or distortion). The second filter to consider is determined by the terrestrial background against which the target is viewed. Thirdly, we shall consider the target itself. The more detailed information we require about the particular target per unit time, the more widebanded our radar illuminator must be, and the larger the bandwidth of the propagation medium between this scene to be investigated and the observation platform. We must, therefore, tailor the illuminating waveform so as to obtain adequate information about the object of interest and at the same time ensure minimum adverse influence of the intervening transmission medium and of the background against which the target is viewed.

In order to provide matched illumination in relation to a given target we shall have to structure the illumination/reception system both in time/frequency and in space. Knowing all there is to be known about the target, a matched illumination can be structured such that zero bandwidth is required to recognize the target. All we then lack is one bit of information, namely that the target is there (Gjessing, 1978*a*, 1978*b*).

6   Introduction

From information theory we know that it takes zero bandwidth to detect a sinusoidally varying signal if we know the period and sufficient observation time is provided. We receive, however, no more information. The matched illumination concept is illustrated in Fig. 1.4. Matched illumination gives upon reception a delta function in the frequency domain from the target to which the illumination is matched.

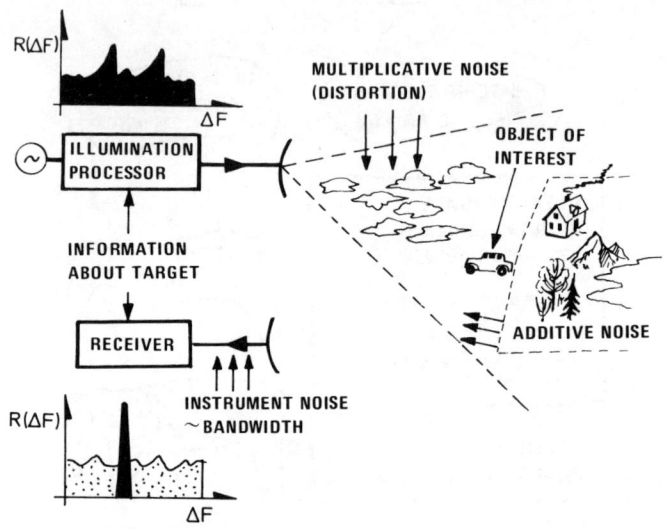

Fig. 1.4   Matched illumination gives a delta function in the frequency domain when scattered back from the object to which the illumination is tailored. (Gjessing, 1981a, © IEEE)

We have at our disposal a radar illuminator which can be structured in the frequency domain. We compose an illuminating frequency spectrum in such a way as to obtain constructive interference by all the reflecting facets of the target. Having structured the illumination in the time domain for optimum coupling to the target, it remains to shape the phase front in space so as to obtain maximum coupling to the particular reflecting structure of interest. By making use of a matrix antenna (two-dimensional broadside array or a SAR) as the radar receiver, the phase and amplitude at each receiver element can be controlled by a computer system so as to provide an antenna system which is matched to the phase front of the wave system which is reflected back from the target of interest, whereas the waves originating from the terrestrial background are suppressed.

The space/frequency system giving information about transverse and lateral distribution of scatterers within a target is illustrated in Fig. 1.5.

If the three signature domains which we have at our disposal (space, frequency and motion pattern) are statistically orthogonal as regards target,

SHAPE AND SIZE OF OBJECTS BY E-M WAVES

A SPACED ANTENNAS

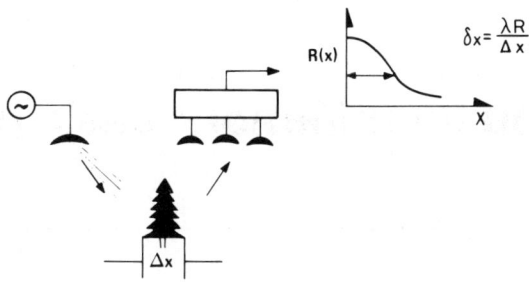

B SEVERAL CORRELATED FREQUENCIES

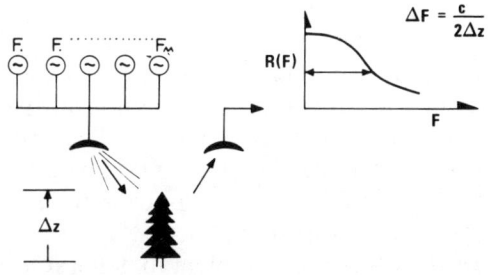

C POLARIZATION

GIVE INFORMATION ABOUT SYMMETRY OF OBJECT

Fig. 1.5 *Spaced antennas (or synthetic aperture) give information about the transverse distribution of the scatterers. The bandwidth of the scattering surface gives information about the distribution in depth (range) of the scattering body, and polarization gives information about the 'symmetry' in the scattering centres*
  a *Spaced antennas*
  b *Several correlated frequencies*
  c *Polarization*

propagation medium and background, we have a simple situation from an information retrieval point of view. We can treat each domain separately and the information processing system (multisensor data fusion) becomes comparatively simple. There are well established algorithms for such data handling. Examples are Kalman filtering, maximum entropy methods, optimum parameter estimation methods etc. If the degree to which the signature domains are orthogonal varies with the conditions prevailing and if also the signatures themselves (target, background, transmission medium) vary rapidly with time, the adaption process becomes increasingly complicated.

Chapter 2

Matched illumination: basic principles

Given all degrees of freedom one can visualize many schemes for enhancing the total scattering from an object of interest at the expense of the contribution from the background against which the target is viewed. In a presentation like the current one, however, where we are facing the rather general problem of matching the illuminator to four dimensions (three in space and one in time), it is of some importance to offer a unified method in the sense that all four dimensions are addressed in the same manner. Seeking a system where the four-dimensional tailoring of the illumination is individual for each dimension, will, at best, complicate the issue conceptually. It is also likely that such particular solutions would prove suboptimal.

Accordingly, we shall solve the problem in general terms, considering the scattering object (surface) to be characterized by a four-dimensional irregularity spectrum (wave-number spectrum).

This approach is based on well-founded generalized Fourier theory such as the classical one by Wiener (1930). This involves describing the scattering surface as decomposed into its Fourier components to which we match individual continuous wave (sine wave) electromagnetic waves.

It takes little imagination to realize that this invovles a superposition of a set of individual phase-locked sine waves (Fourier components) from individual electromagnetic oscillators so as to produce a waveform that is matched to the structure (Fourier components) of the scattering body. Note that for this simple approach to be conceptually meaningful in relation to practical targets, we confine ourselves to a target characterization based on the diffuse back-scattered component. This means that we must illuminate the scattering surface with waves whose wavelength is small compared with the roughness scale of the scattering body.

This wave-number matching will now first be introduced in a qualitative semi-intuitive manner based on simple principles of physics. Afterwards a mathematical approach will be presented leading to mathematical expressions that form the basis for signature calculations.

2.1 Detection/identification of objects by wave-number matching (spatial resolution)

We illuminate the object with a set of electromagnetic waves of different but mutually correlated frequencies. We observe the properties of the back-scattered signal, and from this we can draw very definite conclusions about the distribution of the scatterers constituting the scattering body along the direction of radio wave propagation.

Consider first a simple target consisting of two identical reflection points spaced Δz lonitudinally along the direction of wave propagation. A wide spectrum of frequencies is illuminating this object, and the set of interfering waves scattered back from the object is analysed (inversion techniques). Consider two illuminating waves. As the difference in frequency between the two waves increases, a situation arises where the waves reflected from the two reflection points arrive in antiphase at the receiving site, causing a minimum in the interferogram. If the separation between the two reflecting elements is Δz, the phase angle between the waves reflected from the two reflectors is given by

$$\phi = \frac{2\pi(2\Delta z)}{\lambda_1} = \frac{2\pi(2\Delta z F_1)}{c}$$

For frequency F_1 (wavelength λ_1). Changing the phase angle ϕ through 180° so as to move from constructive interference to destructive interference, requries a frequency change to F_2 such that

$$\pi + \phi = \frac{2\pi(2\Delta z F_2)}{c}$$

The frequency difference between two successive interference minima is thus given by

$$F_2 - F_1 = \Delta F = \frac{c}{2\Delta z}$$

This tells us that by illuminating an object of size Δz, with a set of electromagnetic waves having different frequency, the frequency difference between a situation leading to constructive interference and one leading to destructive interference is given directly as the ratio $c/4\Delta z$. If we wish to resolve a structure within the scattering object of longitudinal extent Δz, we need an illumination with bandwidth $\Delta F = c/2\Delta z$. Fig. 2.2 illustrates this; constructive interference is achieved from an object at distance z_0 if the object is illuminated by two waves with frequency difference $\Delta F = c/2z_0$. By the same token, scattering centres within the target may contribute constructively if the two illuminating waves have a frequency difference $\Delta F = c/2\Delta z$, depending upon the exact range to the target. With a radially moving target this shows up as maximum-depth modulation in the difference frequency return.

10 Matched illumination: basic principles

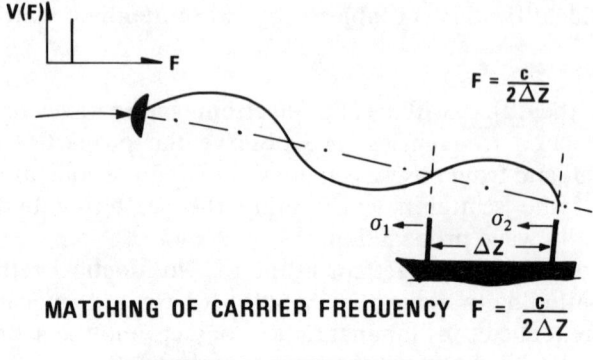

MATCHING OF CARRIER FREQUENCY $F = \dfrac{c}{2\Delta z}$

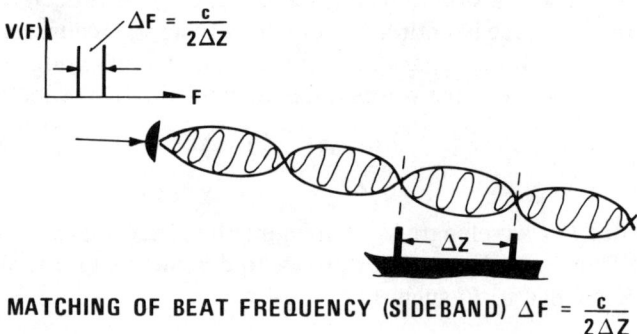

MATCHING OF BEAT FREQUENCY (SIDEBAND) $\Delta F = \dfrac{c}{2\Delta z}$

Fig. 2.1 Choosing a carrier frequency given by $F = c/2\Delta z$ gives constructive interference from two scattering centres spaced Δz apart. In exactly the same way a beat frequency pattern can be matched to the reflecting structure by choosing two illuminating sine waves with spacing $\Delta F = c/2\Delta z$, i.e. choosing a ΔK of the electromagnetic field which is matched to a wave number of the scattering structure. If we are dealing with a complicated reflecting object characterized by several pairs of scattering centres (several Fourier components) we need the same number of frequency pairs for the ΔK matching or several frequencies for the K matching

Fig. 2.1 illustrates the matching of the carrier frequency itself ($F = c/2\Delta z$) and also the case described above with beat frequency matching. This is illustrated in Fig. 2.2 where we make use of several frequency pairs so as to match the illumination to several ΔK components.

If we have at our disposal several frequency generators and wish to measure range accurately (i.e. produce a radar fence), then obviously all the illuminators will have to be tuned synchronously as in a multicavity waveguide filter. This is illustrated in Fig. 2.3.

Now let us express the problem in mathematical terms. We confine ourselves to a scattering body which is at a distance from the illuminator such

Matched illumination: basic principles

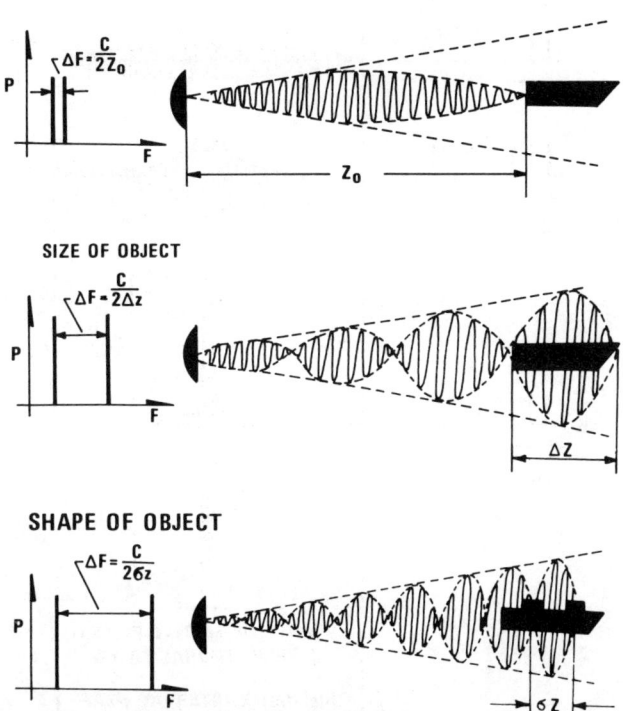

Fig. 2.2 The distance to the object z_0 is determined by transmitting two microwave frequencies with spacing $c/2z_0$; the size of the object is obtained in the same manner by transmitting two frequencies with spacing $c/2\Delta z$. Finally, the shape is obtained by transmitting waves with the larger frequency separation $c/2\delta z$ (Gjessing, 1981a, © IEEE)

that the magnitude of the illuminating field is constant over the scattering body (i.e. $^3\sqrt{\text{volume}} \ll$ range R) and such that the transverse extent of the body is small compared with the first Fresnel zone, i.e. ΔY and $\Delta X \ll \sqrt{(R\lambda)}$ when R is the distance to the object from the illuminator (see Stratton, 1941).

We combine the various factors contributing to the scattered field into one; namely, one which is directly related to the scattering cross-section as a function of distance. We define the function $f(z)$ as the *delay function*. This has dimension field strength per unit length such that the scattering cross-section as a function of distance z along the direction of propagation is obtained by squaring the $f(z)$ function. It is readily seen (Gjessing 1978a) and is intuitively rather obvious that the field strength of the back-scattered wave from an object characterized by the delay function $f(z)$ is given by

$$V(\omega/c) \sim E(\mathbf{K}) \sim \int f(z)\,e^{-j\mathbf{k}\cdot z}\,d^3z \tag{2.1}$$

Matched illumination: basic principles

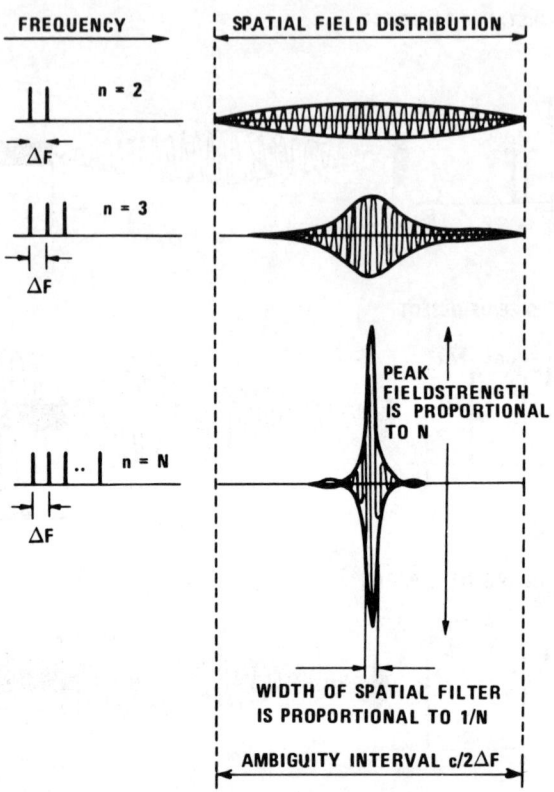

Fig. 2.3 The more frequencies that we have at our disposal for synchronous tuning, the more accurately can we measure distance to the object (Gjessing, Hjelmstad, and Lund, 1982, © IEEE)

Here K is the wave number difference between the scattered and the incident wave (Ratcliffe, 1965; Batchelor, 1955; Megaw, 1957)

$$K = k_1 - k_s$$
$$|K| = 4\pi/\lambda \sin \theta/2$$

The physics of this expression is visualized in Fig. 2.4. Note that the scattering cross-section σ_E is expressed on an amplitude basis rather than in terms of power as is normally the case. Note also that the delay function $f(z)$ in eqn. 2.1 is the same as $\sigma(r)$ in the figure.

As we have already inferred, by introducing the fourth dimension, namely time, the general object is in motion. If we are dealing with translatory motion of a rigid object, then all the Fourier components $\Sigma A(K)$ move with the same velocity such that each illuminating frequency K is subjected to a Doppler shift $\omega = K \cdot V$ where V is the velocity of the object.

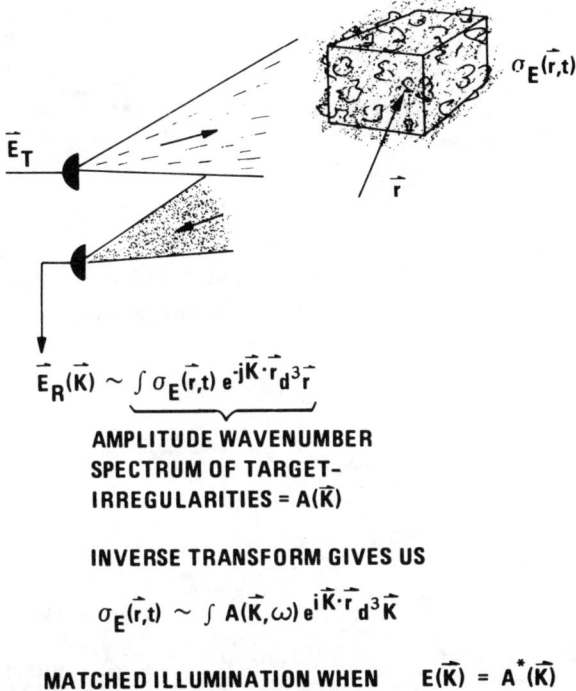

Fig. 2.4 *Any scattering object characterized by a distribution in space of the scatterers ($\sigma_E(r)$) can be decomposed into its Fourier components A(K). We will have constructive interference from all these scatterers if we choose a combination of illuminating sine waves $E_T(K)$ that couple to each of the Fourier components A(K) of the scattering object*

The general object, however, is flexible or compressible such that each Fourier component may have a specific and independent motion pattern. This situation is illustrated in Fig. 2.5.

Our aim now is to derive simple expression for the correlation (in the frequency domain) between two electromagnetic waves at wave numbers $E(K)$ and $E(K + \Delta K)$ scattered back from the sea surface which is characterized by the delay function $f(z)$.

This obviously is a four-dimensional problem (three dimensions in space, one in time). The normalized correlation in ΔK then is written in the conventional manner

$$R(\Delta K, t) = \frac{\langle E(K,t) E^*\{(K + \Delta K), t\}\rangle}{\langle |E(K,t)|^2\rangle} \tag{2.2}$$

Note that $E(K)$ and indeed also $E(t)$ are stochastic functions. This means that if we consider a frozen sea surface structure, and thus for the time being leave t as

14   *Matched illumination: basic principles*

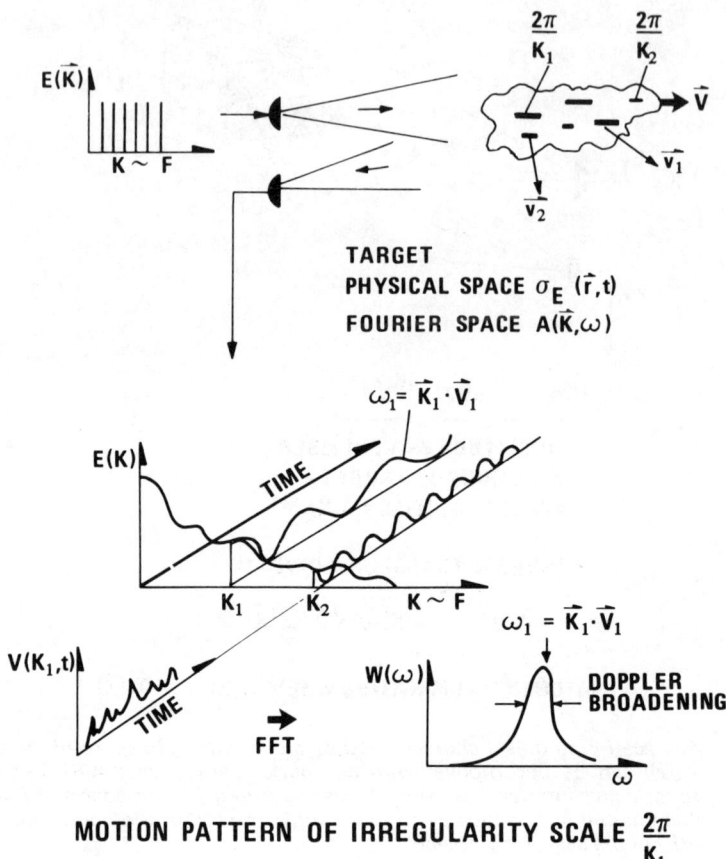

Fig. 2.5   *For a general flexible object, each scattering element or Fourier component may have a specific, independent motion pattern. However, in general the power spectra (temporal Fourier transform) of the individual E(**K**,ω) components may be different and independent of each other*

a constant, which is perfectly acceptable, we shall still have to calculate $R(\Delta K)$ for a large number of $E(K)$ realizations since $E(K)$ is non-deterministic.

If the sea structure were treated as frozen (as the lunar surface investigated by Hagfors, 1961), we should have to obtain an ensemble of realizations by looking at various geographical samples of the sea surface. This, of course, assumes spatial stationarity. If the statistics of the sea is ergodic as well as stationary we obviously can obtain a meaningful description of $E(K,t)$ from time averaging.

We are, however, interested in the spatial delay function $f(z)$ as well as in the motion $f(t)$. A convenient manner in which to measure the velocity distribution of a given spatial scale ΔK is to compute the power spectrum (temporal Fourier transform) of the spatial correlation function $R(\Delta K)$.

To continue where we left at eqn. 2.2, therefore, we compute the temporal Fourier transform

$$FT_t R(\Delta K,t) = FT_t \left(\frac{\langle E(K,t)\, E^*\{(K+\Delta K),t\}\rangle}{\langle |E(K,t)|^2\rangle} \right)$$

The calculation scheme is shown in Fig. 5.23. Measuring the quantity $V^4(f)$ corresponding to the resonance peak of the Doppler spectrum we obtain information about the intensity of the coherent (gravity wave) motion of the sea surface. Measuring the spectral intensity $V^4(f)$ of the skirts in the Doppler spectrum, we obtain information about the velocity distribution of the non-coherent turbulent sea surface irregularities.

After this rather detailed introduction, hopefully it is obvious to the reader that we can leave out the time domain (temporal Fourier transformation) until we have established the appropriate spatial relationships through the spatial correlation $R(\Delta K)$.

Hence, to simplify the concepts, and indeed the mathematics, we shall confine the discussion to the $R(\Delta K)$ function, bearing in mind that we actually calculate the temporal power spectrum of this spatial correlation.

Using then k as the free variable, the nominator can be assumed to be a fixed normalization factor. Substituting eqn. 2.1 in eqn. 2.2, we obtain

$$R(\Delta K) \sim \langle \int f^*(z)\, e^{jK\cdot z}\, dz \int f(z')\, e^{-j(K+\Delta K)z'}\, dz' \rangle$$

It can now be proved that the $R(\Delta k)$ function is a direct Fourier pair of the mean quadratic scattering function $f(z)$. We will here follow a scheme suggested by Jacobsen and Eltoft (see Jacobsen, 1985)

Assuming that the medium is statistically locally homogeneous, by ensemble averaging of the stochastically variables $f(z)$ and $f(z')$ we can rewrite the above equation as

$$R(\Delta K) \sim \iint \langle |f(r)|^2\rangle\, R(z-z')\, e^{-j(K+\Delta K)z'}\, e^{jKz}\, dz\, dz' \qquad (2.3)$$

where

$$r = 1/2(z+z')$$

The autocorrelation $R(z-z')$ is normalized with the mean quadratic scattering function

$$\langle |f(r)|^2\rangle$$

If we now assume that the correlation distance in space is short, we can approximate the autocorrelation function $R(z-z')$ with a delta function $\delta(z-z')$. Then eqn. 2.3 takes the form

$$R(\Delta K) \sim \iint \langle |f(r)|^2\rangle\, \delta(z-z')\, e^{jKz}\, dz\, e^{-j(K+\Delta K)z'}\, dx'$$

Solving the inner integral, we obtain

$$R(\Delta K) \sim \int \langle |f(z)|^2\rangle\, e^{j\Delta Kz}\, dz$$

Using the dispersion relation for the modulated component of the product for the one-dimensional case:

$$2\pi \, \Delta F = c\Delta k$$

This gives the equation

$$R(\Delta F) \sim \int R(r) \, e^{-j2\pi \Delta F c^{-1} r} \, dr \tag{2.4}$$

where

$$R(r) = \langle |f(z)|^2 \rangle$$

Eqn. 2.4 states that the envelope of the complex correlation in the frequency domain of waves scattered back from a surface is given by the autocorrelation function $R(r)$ characterizing the surface.

Turning back now to our practical computing problem, we again note that we are dealing with a sea surface which is in motion characterized by a coherent wave giving a resonance in the Doppler spectrum due to the ordered velocity and a broad skirt arising from the wide velocity distribution of the incoherent (turbulent) components (see Chapter 5).

Hence, as depicted in Fig. 5.23, we compute the temporal Fourier transform of the spatial correlation, also obtaining information about the motion pattern

$$FT_t\{R(\Delta F, t)\} \sim FT_t\{\int R(r,t) \, e^{-j2\pi \Delta F c^{-1} r} \, dr\}$$

The equation states that the envelope of the complex correlation in the frequency domain of waves scattered back from a surface is given by the autocorrelation function $R(r)$ characterizing the surface (see also Gjessing, Hjelmstad, and Lund 1982; Jacobsen 1985). Fig. 2.6 summarizes the physics of eqn. 2.4.

Note that in order to obtain a unique description of the scattering object with one carrier frequency F only, we must assume that the individual scattering centres are sufficiently broadbanded for the scattering cross-section to be constant to the first order over a certain range of F. This range obviously will be large in comparison with the maximum value of ΔF.

In practice, what contributes to the scattering cross-section of dominating scattering centres are corner-reflector type structures, which in practice are very broadbanded. Scattering centres composed of high Q-factor resonators hardly exist in practice.

Finally, Fig. 2.7 gives multifrequency signatures of some simple objects in the form of two delta functions whose Fourier transform is a cosine function, a Gaussian distribution of the scatterers etc.

Note that if the object has scatterers that are distributed in a periodic fashion, then the $R(\Delta F)$ function peaks at a value of ΔF corresponding to the period δz of the function. Hence $R(\Delta F)$ is a maximum when $F = C/2\delta z$ whereas the width of the $R(\Delta F)$ function is determined by the degree of truncation of the periodic delay function.

Matched illumination: basic principles

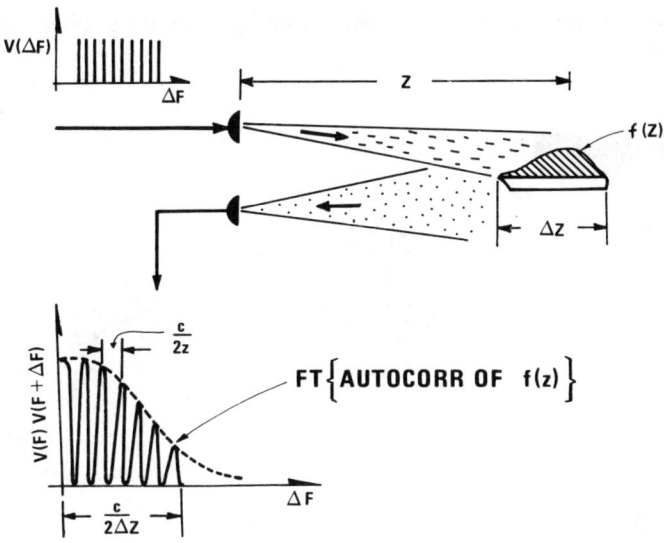

Fig. 2.6 *Modulus of the $R(\Delta F)$ correlation is the Fourier transform of the target autocorrelation function. The oscillations give information about range to the object*

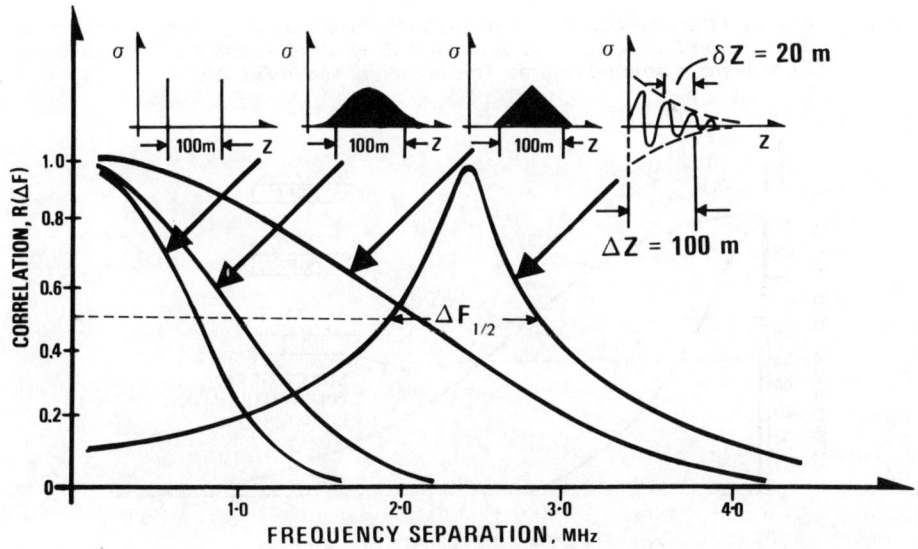

Fig. 2.7 *Knowing the distribution in depth (range) of the scattering elements (centres), we can calculate the radar signature. Hence we know the composition of sine waves with which we will illuminate the object to achieve matched illumination, i.e. constructive interference from all the scattering elements*

18 Matched illumination: basic principles

In summarizing this section on wave-number matching, Figs. 2.8 and 2.9 are presented.

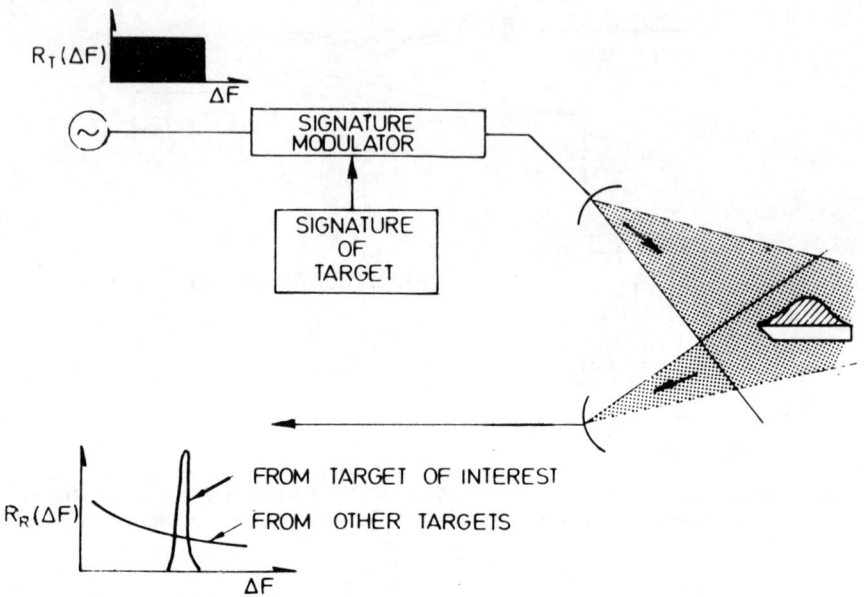

Fig. 2.8 Knowing the distribution in depth (range) of the scattering centres, a matched illumination function can be structured so as to get constructive interference from all the scattering centres. This is exemplified in Fig. 2.9.

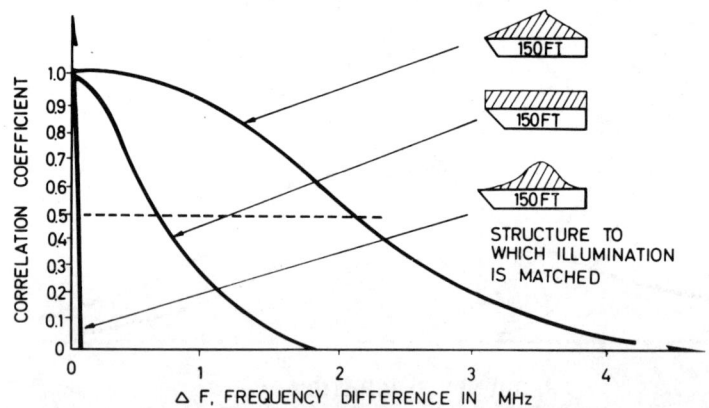

Fig. 2.9 Theoretical results of matched illumination. When everything is known about the object of interest, zero bandwidth is needed for recognition (detection) and only one bit of information is obtained, namely that there exists such an object in the radar beam (Gjessing, 1978a)

Matched illumination: basic principles

To get some feeling for the relative merits of a matched illumination radar and one based on conventional concepts, an example involving typical numbers is presented in the following:

Multifrequency radar versus pulse radar

Example:
Contributing scattering centres N of target is 100
Smallest feature (distance between scattering elements) is 15 cm

Pulse radar:
Pulse width required is 1 ns
Bandwidth required (noise and signal) $\Delta F = 1$ GHz
PRF 1 KHz give 150 km ambiguity interval
Integration period $\tau = 100$ ms
m = 100 pulses during this integration period

Multifrequency radar:
100 beat-frequencies required to resolve 100 scattering centres (scales)
With n frequencies we get $1/2(n-1)n$ beat channels. We thus require $n = 15$ frequencies
Full parallel processing with Doppler filter bandwidth $\Delta f = 10$ Hz
For a given target range, target shape, antenna gain, noise factor and peak power the ratio of S/N for the two radar types becomes

$$\frac{(S/N)_{\text{multifrequency}}}{(S/N)_{\text{pulse}}} = \frac{\Delta F/m}{n\Delta f} N^{\frac{1}{2}}$$

First factor is the bonus of the effective noise bandwith
The second is bonus from matching the radar to the target, for the given example with
$\Delta F = 1$ GHz
$m = 100$ pulses during integration interval
$n = 15$ radar frequencies
$\Delta f = 10$ Hz Doppler filter bandwidth

$$\frac{(S/M)_{\text{multifrequency}}}{(S/N)_{\text{pulse}}} \approx 70 \text{ dB}$$

Before we leave this section on ΔK matching (microwave holographic methods), let us, for the purpose of ensuring a thorough physical understanding, adopt the more direct approach of a communications engineer who requires information about circuit bandwidth.

Consider the case of diffuse (not specular) reflection (scattering) from a rough surface. The surface consists of a collection of scattering elements distributed in depth z from which a set of waves is reflected back to the receiver. These waves will interfere (multipath fading) and the result is a signal with limited bandwidth, i.e. with limited correlation properties in the frequency domain. We shall now calculate this correlation function (bandwidth of reflector).

To avoid the confusion which often arises when the term 'bandwidth' is used in relation to scattering processes, let us define what we mean by bandwidth. Consider the case where several radio waves having different frequencies are transmitted simultaneously. At the receiver, the power at each of these frequencies is measured as a function of time (i.e. we measure $P_{F_1}(t)$, $P_{F_2}(t)$ etc). If we take the instantaneous ratio of power at the various frequencies and integrate the ratio (i.e we form $\int \{P_{F_1}(t)\}/\{P_{F_2}(t)\} \, dt$) then we obtain information about bandwith.

On the other hand, if we integrate the signal at either frequency over the appropriate time interval by forming $\int\{P_{F_1}(t) \, dt\}/\int\{P_{F_2}(t) \, dt\}$ we do not obtain information about bandwidth but something which is often referred to in radio science literature as 'the wavelength dependence of the scatter circuit'. Forming the $\int \{P_{F_1}(t)\}/\{P_{F_2}(t)\} \, dt$ is one way of obtaining information about bandwidth. Another is the following.

As in the case above, we transmit a set of radio waves having different frequencies. We now make sure that all these frequencies are correlated in amplitude and phase. This is, as an example, achieved by amplitude modulating a carrier, thus obtaining two sidebands $2F_{AM}$ apart, F_{AM} being the frequency of the modulating wave. These sidebands, obviously, are correlated in amplitude and phase. At the receiving end we pick up the two sidebands and correlate one with the other (i.e. we form the cross-correlation function $R_{12}(\tau)$). The more narrowbanded the transmission channel, the poorer is the correlation (by transmitting many correlated waves spread over a frequency band, we can find the complete autocorrelation function $R(\Delta F)$ in the frequency domain) (Hagfors, 1959; Waterman et al., 1961; Crawford et al., 1959).

If the scattering cross-section σ associated with a scattering element distance z from the receiver is denoted $\sigma(z)$, the associated delay function of the received field is

$$P(\tau) \simeq \sigma(z/c)$$

where c is the velocity of light.

Using analogous results from network theory, we have the following relationships. The Fourier transform (FT) of the delay function is known as the transfer function $F(\omega)$. Transfer function is thus

$$F(\omega) = FT\{P(\tau)\}$$

Furthermore, the resulting power spectrum $W(\omega)$ is given by

$$W(\omega) = F(\omega)\, F^*(\omega)$$

Thus, knowing the distribution in depth of the scattering elements $\sigma(z)$, the bandwidth of the reflecting surface $(\Delta\omega)$ given by $W(\Delta\omega)/W(0) = 1/2$ is obtained by a simple Fourier transformation of the $\sigma(z/c)$ function (Gjessing, 1973).

To illustrate the essential points, let us discuss as an example the scattering process from vegetation. We shall assume that the ground surface being illuminated consists essentially of coniferous trees having needles distributed evenly in depth in such a way that the shadowing effect becomes progressively more dominant as the wave progresses. Let us, for the sake of simplicity, assume an exponential shadowing effect such that the illuminated scattering facets are distributed in an exponential manner in depth. This leads to a set of waves arriving at the receiver. These waves will interfere and the result is a limited correlation bandwidth of the reflecting surface. We shall now calculate this bandwidth.

The delay function of these reflected waves is given by

$$P(\tau) = e^{-\alpha\tau}$$

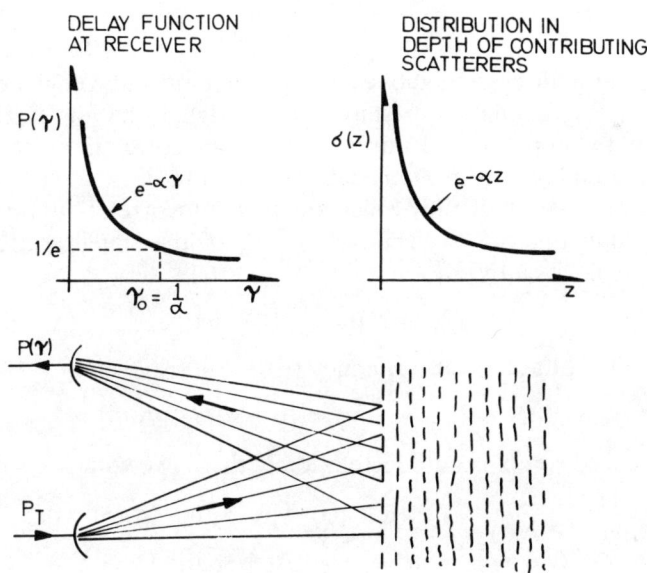

Fig. 2.10 *Geometry of the scattering process (correlation properties in the frequency domain)*

since the distribution in depth of scattering cross-section is assumed to be exponential. The 1/e width of this delay function is given by

$$\tau_0 = \frac{1}{\alpha}$$

The Fourier transform (*FT*) of the delay function is given by

$$F(\omega) = FT \text{ (delay function)}$$
$$= FT(e^{-\alpha\tau})$$
$$= (\alpha + j\omega)^{-1}$$

The resulting power spectrum is thus (van Trees, 1971)

$$W(\omega) = f(\omega) F^*(\omega) = (\alpha^2 + \omega^2)^{-1}$$

The half power width $\Delta\omega$ of this spectrum is given by

$$\frac{W(\omega)}{W(0)} = \frac{1}{2} = \frac{\alpha^2}{\alpha^2 + \Delta\omega^2}$$

i.e.
$$\Delta\omega = \alpha = 1/\tau_0$$

Hence, if the 1/e penetration depth of the electromagnetic wave is $c\tau_0 = \Delta z$ metres, the 'bandwidth' (width of power spectrum) is given by

$$\Delta F = \frac{c}{2\pi\Delta c}$$

The 'bandwidth' discussion above may appeal to the radio engineer. However, to the scientist familiar with statistics and signal analysis it may be more meaningful to express the results in terms of the correlation properties of the signal scattered back from the rough terrain surface.

Specifically, we shall now calculate the autocorrelation function in the frequency domain $R(\Delta\omega)$. The voltage V_1 of the signal scattered back at frequency ω is given by

$$V_1 = F(\omega) = (\alpha + j\omega)^{-1}$$

Similarly, the voltage V_2 at frequency $(\omega + \Delta\omega)$ is given by

$$V_2 = F(\omega + \Delta\omega) = [\alpha + j(\omega + \Delta\omega)]^{-1}$$

The normalized complex autocorrelation of these two voltages is then given by

$$R(\Delta\omega) = \frac{{}_\infty\int^\infty (\alpha + j\omega)^{-1} [\alpha - j(\omega + \Delta\omega)]^{-1} d\omega}{{}_\infty\int^\infty (\alpha^2 + \omega^2)^{-1} d\omega}$$

Solving this integral we derive the following expression for the modulus of the autocorrelation function

$$R(\Delta\omega) = [1 + (\Delta\omega/2\alpha)^2]^{-\frac{1}{2}}$$

Defining now the correlation distance in the frequency domain to be the half width of the autocorrelation function, we have

$$\Delta\omega_{\frac{1}{2}} = 2\alpha \sqrt{3}$$

$$\Delta F = \frac{\sqrt{3}}{\pi} \frac{c}{\Delta z} \text{ Hz}$$

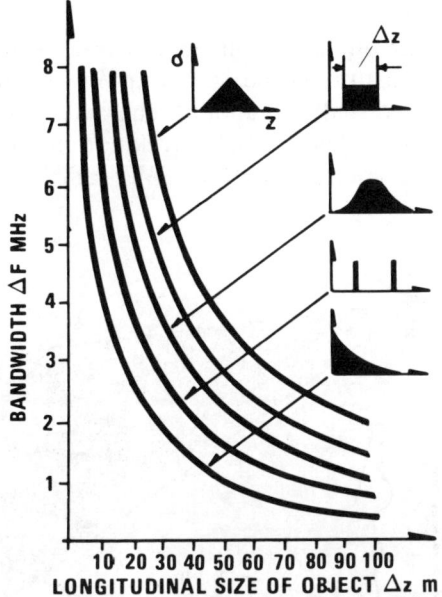

Fig. 2.11 *'Bandwidth' of a reflector consisting of vegetation with exponential distribution in depth of the scattering elements (exponential shadowing effect)*

For easy visualization, this simple expression is plotted linearly in Fig. 2.11. Note that the distribution in depth of the scatterers is assumed to be exponential, such that the height of the vegetation shown as the abscissa of Fig. 2.11 is the 1/e width of the exponential depth distribution function. Had we assumed a different distribution in depth of the vegetation, we would have obtained a different curve relating bandwidth and height of vegetation. However, the inverse relationship would always exist, whereas the constant of proportionality would differ (Gjessing, 1978a).

The relationship between broadband radar and communication systems is treated by Hjelmstad (1983), and generalized schemes for matched spread spectrum communication systems is considered in Hjelmstad and Skaug (1985) (see also Hjelmstad, 1983).

2.2 Transverse distribution of scattering elements: spatial correlation of scattered waves, synthetic aperture method

We have now discussed the correlation properties of electromagnetic waves with different frequency, and have seen that by measuring the degree to which waves of different frequency are correlated, we obtain information about the longitudinal distribution of the scatterers. If we are dealing with a thin reflector (zero distribution in depth) then the bandwidth of the reflector is very large. Conversely, if the scatterers are distributed over a large region in space, the bandwidth is small.

We shall now focus our attention on the transverse distribution of the scattering elements constituting the scattering object. In order to reveal this transverse structure, we shall use another characteristic property of electromagnetic waves, namely, their spatial correlation properties. This forms the basis for the synthetic aperture method. We shall illuminate the scattering

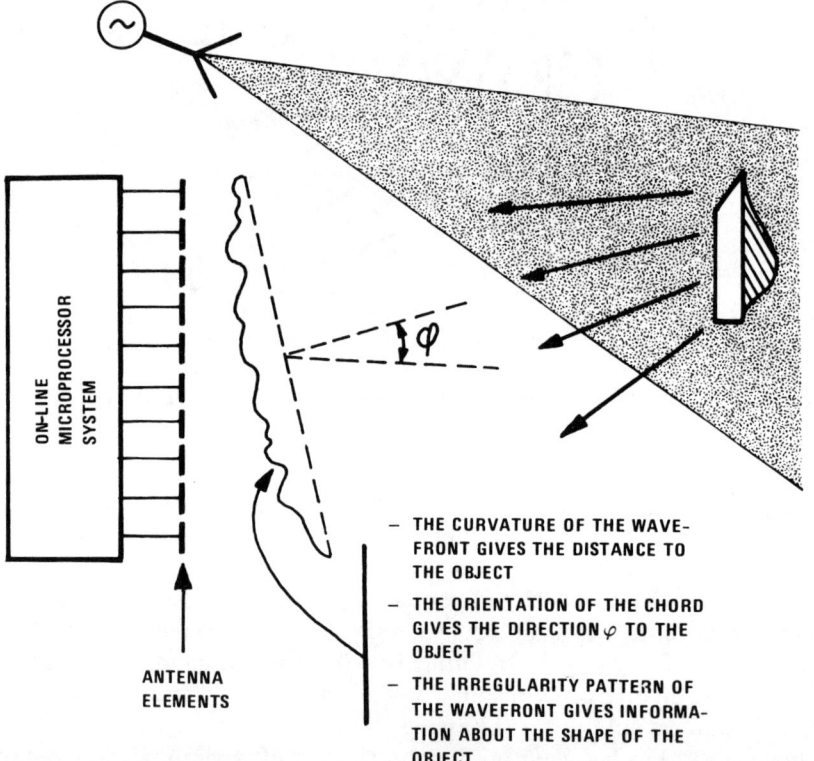

- THE CURVATURE OF THE WAVEFRONT GIVES THE DISTANCE TO THE OBJECT
- THE ORIENTATION OF THE CHORD GIVES THE DIRECTION φ TO THE OBJECT
- THE IRREGULARITY PATTERN OF THE WAVEFRONT GIVES INFORMATION ABOUT THE SHAPE OF THE OBJECT

Fig. 2.12 *Measuring the detailed structure of the phase front of the backscattered waves gives us information about position of object, size and shape. The significance of this statement is illustrated in Fig. 2.13*

object with one single frequency, and at the receiving site we shall make use of a set of antenna elements distributed along a base line or in a plane which is perpendicular to the line joining the receiving array and the scattering object. At each element we shall measure the amplitude and the phase of the impinging wave. On the basis of these point observations of field strength, information about the target can be extracted. This will be the subject of the current section.

First, consider Fig. 2.12, illustrating the scattering process. We illuminate a surface with a single wave. Depending on the properties of the surface, we obtain a certain field-strength distribution of the scattered field $E_s(x)$.

If the scattering object is a point target, the phase front of the scattered wave will be a smooth sphere. Measuring the radius of this sphere gives us information about the distance to the target. The direction to the target is obtained by the orientation of the chord as indicated in Fig. 2.12. If the target is distributed in a direction orthogonal to that of the illuminating radio beam, we shall have a set of spherical waves giving a spatial interference pattern.

The basic principle of spatial interferometry (synthetic aperture antenna) is further illustrated in Fig. 2.13. Measuring the transverse spatial distribution of field strength gives us information about transverse shape and size of the scattering object if the range is known. Flying with a small side-looking antenna through this interference pattern reveals the target structure (see later), as illustrated in Fig. 2.14 (Applebaum, 1976; Gabriel, 1976; Brown *et al.*, 1969).

Having introduced this synthetic aperture concept in terms of first-principle physics, we shall now calculate the angular distribution $P(\theta)$ scattered from a

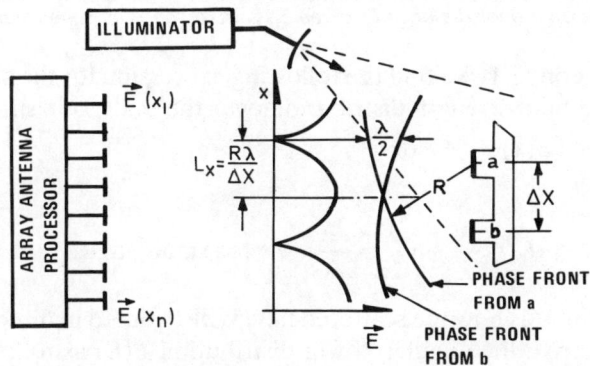

Fig. 2.13 *Transverse spatial interference pattern gives information about transverse shape and size of the scattering object provided we know the position of the object (see later)*

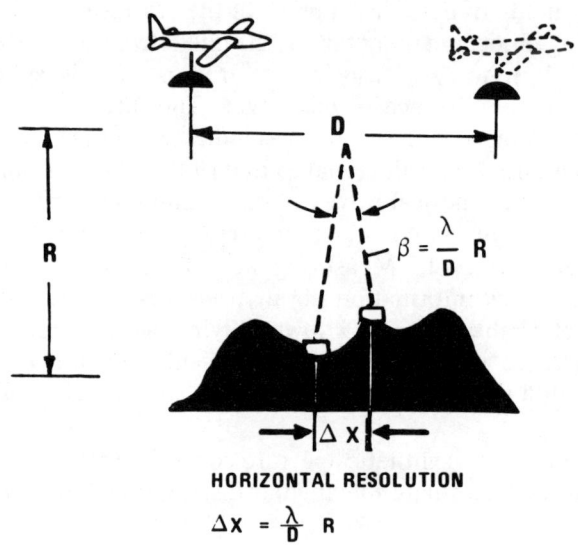

Fig. 2.14 *Measuring with a small antenna the field strength as a function of transverse position x over distance D, gives a synthetic angular resolution β = λ/D*

surface. From eqn. 2.1, we find the following expression for the scattered field in terms of the field-strength distribution over the scattering surface:

$$E_s(K) \sim \int E(x) \, e^{-jK \cdot x} \, dx \qquad (2.5)$$

Since

$$K = \frac{4\pi}{\lambda} \sin \frac{\theta}{2} \simeq \frac{2\pi}{\lambda} \theta \quad \text{for small angles}$$

This equation tells us how the scattered field is distributed in direction θ. From this we shall derive the angular power distribution $P(K)$ as follows (Gjessing and Boerresen, 1968):

$$\begin{aligned} P(K) &\sim E_s(K) \, E_s^*(K) \\ &\sim \iint E^*(x) \, E(x + r) \, e^{jK \cdot x} \, e^{-jK(x + r)} \, d^3x \, d^3r \\ &\sim \int d^3r \, e^{-jK \cdot r} \int d^3x \, E(x) \, E(x + r) \end{aligned}$$

The second integral is immediately recognized as the spatial autocorrelation $R_E(r)$ of the field-strength distribution.

Hence

$$P(\mathbf{K}) \sim P(\sin \theta) \sim \int R_E(r)\, e^{-j\mathbf{K}\cdot r}\, d^3r \qquad (2.6)$$

This equation tells us that the angular power spectrum (radiation pattern) of the scattered wave is the Fourier transform of the field-strength distribution over the scattering region when this distribution is expressed statistically in terms of its spatial autocorrelation, as illustrated in Fig. 2.15.

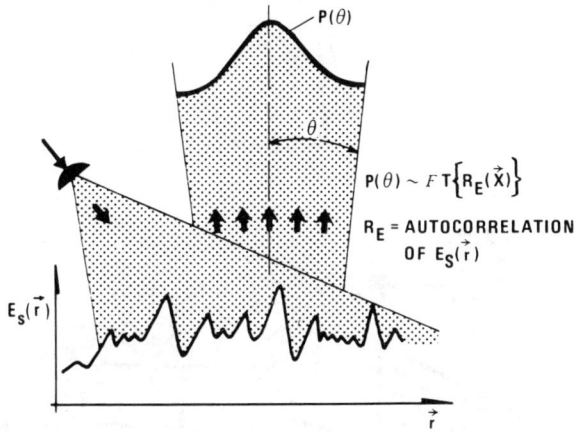

Fig. 2.15   *Geometry of the back-scattering process. The illuminating field gives rise to a scattered field-strength distribution $E_s(x)$*

This is a relationship which is very well known from antenna theory: the radiation pattern (angular power distribution) of an antenna with aperture a is obtained by the Fourier transform of the field-strength distribution over this aperture. Thus, if our $E(x)$ function is a rectangular one, implying that the field strength is evenly distributed over the antenna aperture, then the angular power distribution is of the form $(\sin \theta/\theta)^2$ and the beamwidth $\beta = \lambda/D$ where D is the aperture size. Conversely, if the field-strength distribution over the aperture is of the $\sin x/x$ form, then the angular distribution of the scattered wave is a rectangular one. We shall be using these simple relationships extensively in the subsequent sections.

Now let us return to eqn. 2.6 and use this as the basis for studying another important property of the scattered field, namely the spatial correlation of field strength. In the discussion which we have just completed, we considered the case of a 'transmitting antenna'.

Now let us consider the case of a receiving antenna. Our receiving antenna consists of set of antenna array elements which permits us to measure amplitude and phase at each array element. The power reaching this array

28 Matched illumination: basic principles

antenna is distributed as $P(\theta)$ over an angular region. Applying the inverse Fourier transform of eqn. 2.6, it is intuitively obvious that we obtain information about the spatial correlation properties of the field strength

$$R_E(r) \sim \int P(K) e^{jK \cdot r} dK \tag{2.7}$$

This equation tells us that the spatial correlation of the scattered field is the Fourier transform of the angular power distribution.

We see from Fig. 2.13 that if we are to resolve an object of transverse extent Δx by means of a receiving antenna array at distance R from the object, we shall have to measure the field-strength distribution over a spatial region

$$L_x = \frac{R\lambda}{\Delta x}$$

Summing up these findings, we should note that by measuring the field-strength distribution across a broadside array (amplitude and phase at each

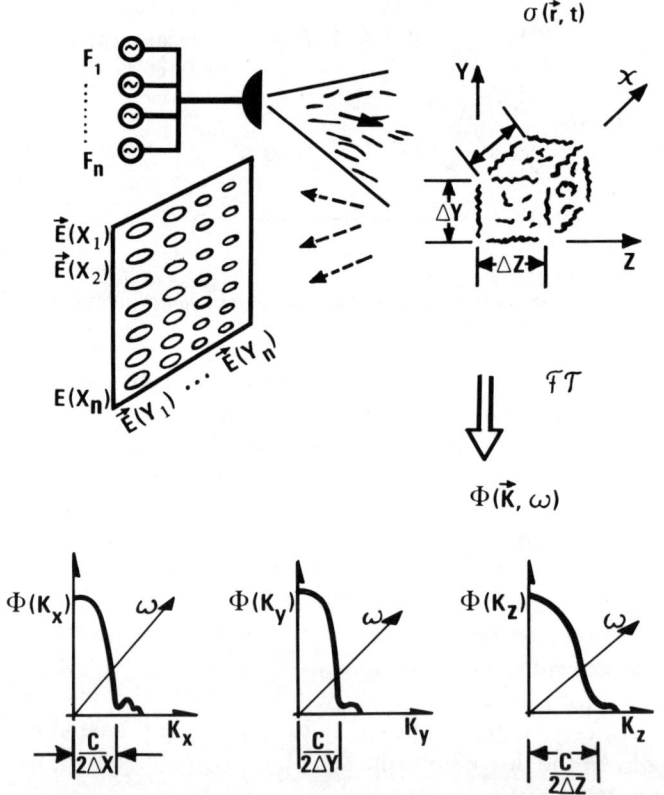

Fig. 2.16 A non-rigid scattering object can be characterized as a four-dimensional irregularity spectrum $\Phi(K,\omega)$. Matched illumination involves producing a 'four-dimensional hologram' by an appropriate time-variable combination of illumination

array point), we obtain direct information about the transverse scattering properties of the scattering object.

It takes little imagination to see the analogy between this spatial autocorrelation function and the autocorrelation function in the frequency domain discussed above. In the multifrequency case, we can 'filter out' certain longitudinal spatial distributions of the scattering object by using frequency filters. In the case of the broadside array or an SAR, we can 'filter out' certain transverse properties of the scattering elements by providing a 'spatial filter'.

We see that with a multifrequency synthetic aperture system we will obtain directly the two-dimensional Fourier transform (the hologram) of the scattering surface. If, by matched illumination, we produce the hologram of the object of interest, we obtain a system with optimum sensitivity and target identification capability. We have then managed to match the illumination to a three-dimensional target irregularity spectrum. Only one dimension then remains, as illustrated in Fig. 2.16, namely that of time, which will be the subject of later sections and in particular of Section 2.4. As an intermediate summing up, however, Fig. 2.16 is presented.

Before we proceed to consider certain practical and illustrative aspects of the multifrequency 'zooming SAR' system, let us sum up the section on three-dimensional matched illumination.

First, consider Fig. 2.17 showing the transverse spatial autocorrelation of various objects observed with a 100 GHz illuminator at 3 km range. Note that the elements in the receiving matrix antenna (see Fig. 2.13) have to be in the order of 2 metres apart in order to distinguish between the 18 m object of different shapes.

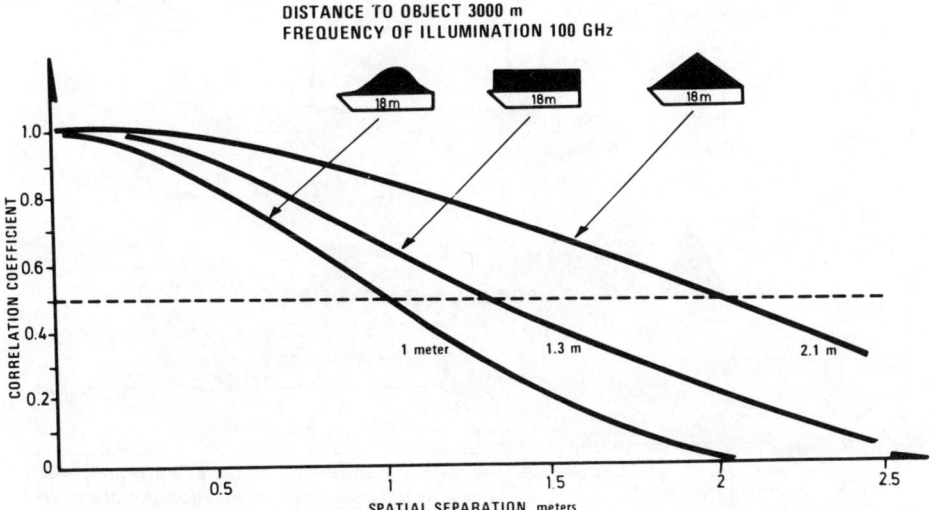

Fig. 2.17 *Theoretical results illustrating the identification potential of a single frequency illuminator and a multi-element (SAR) receiving antenna system*

30 Matched illumination: basic principles

In summing up the last two sections on trasverse resolution and longitudinal resolution, respectively, Figs. 2.18 and 2.19 are presented. From Fig. 2.18 we see that for equal transverse and longitudinal resolution, the following conditions have to be satisfied:

$$\frac{\Delta F}{F} = \frac{D}{2R}$$

Here $\Delta F/F$ is the relative bandwidth, R is target range and D is antenna aperture (SAR integration range). Finally, Fig. 2.19 illustrates the basic physics of the two-dimensional target resolution concept.

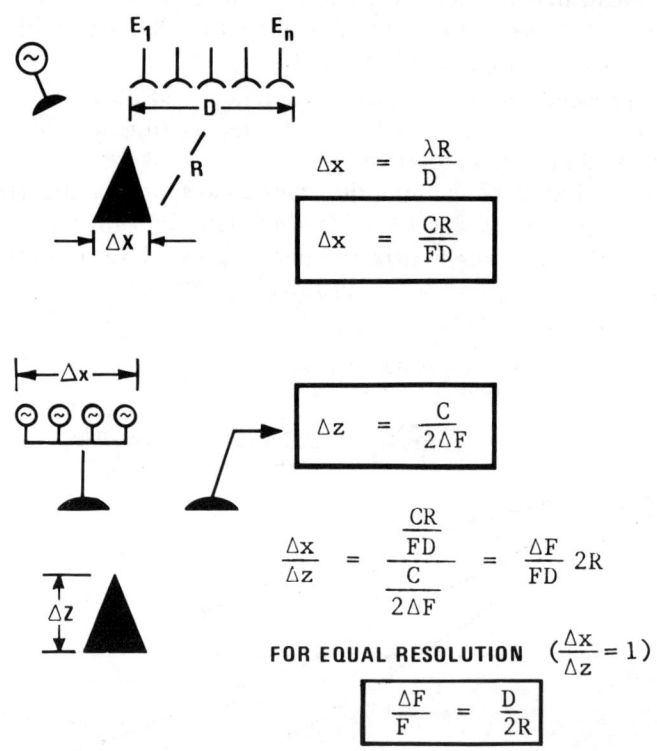

Fig. 2.18 *In order to obtain the same resolution for the two orthogonal directions (longitudinal and transverse), the ratio of bandwidth to carrier frequency must be the same as the ratio of antenna dimension (SAR integration distance) and twice target range*
a Spaced receiving antennas (SAR) give transverse resolution
b Spaced frequencies give resolution in depth

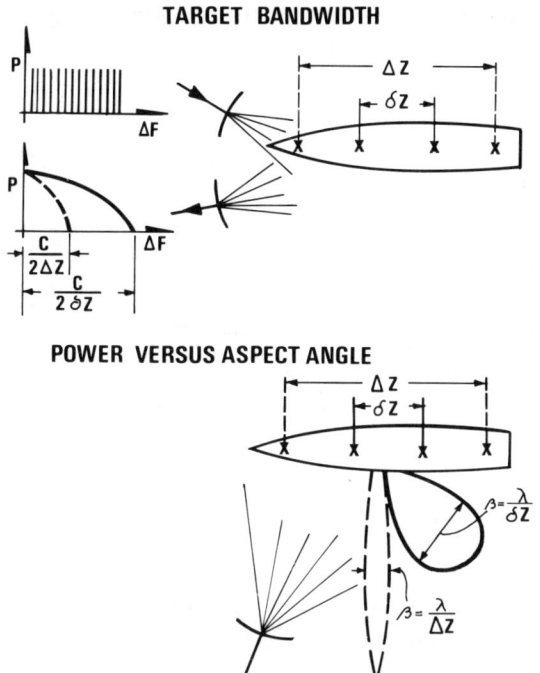

Fig. 2.19 *Measuring target bandwidth and spatial distribution of scattered field gives information of two-dimensional target shape*

Before ending this section on characterization of a scattering object by three-dimensional shape and size, let us briefly use intuitive basic physics concepts for the prupose of illustrating the principle of SAR (see, for example, Beal *et al.*, 1981).

If the objective, as an illustrating example, is an optimum system for detection, identification and acquisition of ships against the sea surface, we are certainly not interested in an image of the background. Neither are we interested in measuring the position of the ship to decimetric accuracy, although we may well wish to resolve the target down to such an accuracy. Suppose that the identification of the target is achieved using the multifrequency mode with resolution down to approximately 20 cm whereas measurement of position is achieved by a course resolution multifrequency mode and by a SAR method. Note that this involves extracting information about the target directly from the raw video signal without transformation to an image.

To ensure a thorough physical understanding of the basic principles, we shall move slowly into the details of the problem. Consider first a point target on the sea surface at a distance R from the radar as indicated in Fig. 2.20. Velocity of the SAR platform is V and the antenna beamwidth is β. The Doppler shift f is given by

$$f = \frac{1}{2\pi} \mathbf{V} \cdot \mathbf{K} \tag{2.8}$$

which we shall use in Section 2.4.

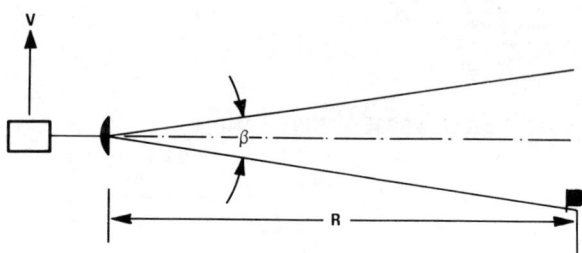

Fig. 2.20 *Geometry of the simple SAR system*

The magnitude of $|\mathbf{K}|$ is $4\pi/\lambda$ and the direction varies from $-\beta/2$ to $+\beta/2$. Hence, when the antenna beam sweeps over the point target the Doppler frequency varies from

$$f_{\text{MIN}} = -\frac{2V}{\lambda}\sin \beta/2$$

to

$$f_{\text{MAX}} = \frac{2V}{\lambda}\sin \beta/2$$

For narrow beams, therefore

$$\Delta f = \frac{2V}{\lambda}\beta$$

Example:
$$\lambda = 5\,\text{cm}$$
$$V = 200\,\text{m/s}$$
$$\beta = 10°$$

i.e.
$$\Delta f = \pm 700\,\text{Hz}$$

Obviously, by the same set of arguments, the Doppler spectrum resulting from a sea surface (neglecting the wave velocity) has a width

$$\Delta f = \frac{2V}{\lambda}\beta$$

The observation time for a point target at range R is

$$T = \frac{\beta R}{V}$$

Thus, as an example, if the range is 200 km and β and V are 10° and 200 m/s, respectively, as in the example above

$$T = \pm 87 \text{ s}$$

Then consider a target having two reflectors spaced Δx apart as illustrated in Fig. 2.13. The interference pattern at distance R from the target will have a lobe structure with lobe width

$$L_x = \frac{R\lambda}{\Delta x}$$

Flying the antenna through this lobe structure produces a scintillation period

$$t = \frac{\lambda R}{\Delta x} / V$$

and a scintillation frequency

$$f = \frac{V \Delta x}{\lambda R}$$

For a 100 m target therefore we shall experience a scintillation frequency of 0·5 Hz.

If there are many scattering elements contributing within the target, the overall size of which is Δx, these will produce scintillations with frequencies less than f.

Let us now consider a general target where the scattering cross-section is distributed over the target as $\sigma(x)$, as shown in Fig. 2.21.

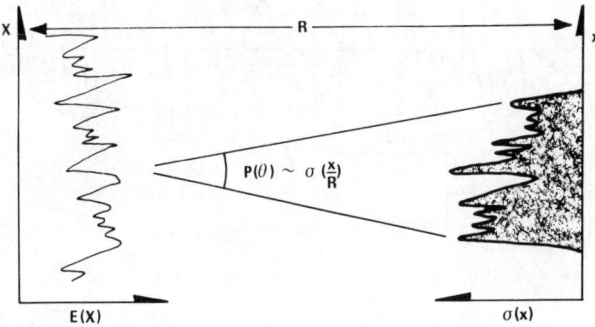

Fig. 2.21 Geometry of the scattering process

Matched illumination: basic principles

We have already seen that the angular amplitude spectrum $E_s(K)$ of the scattered wave is the Fourier transform of the spatial field-strength distribution $E(x)$ as seen from fig. 2.15, i.e.

$$E(K) \sim \int E(x)\, e^{-jK \cdot x}\, dx$$

We also developed the expression for the angular power spectrum $P(\theta)$ by forming the product $E(K)\, E^*(K)$. Hence

$$P(\theta) = E(K)\, E^*(K)$$
$$\sim \int R_E(r)\, e^{-jK \cdot r}\, dr$$

where $R_E(r)$ is the autocorrelation function of the field strength interferogram $E(x)$ and where r is an increment in x.

$$|K| = \frac{4\pi}{\lambda} \sin \theta/2$$

Denoting again the velocity of the SAR antenna by V, we have

$$r = V\psi$$

where ψ is an increment in time

$$K = \frac{\lambda}{c}$$

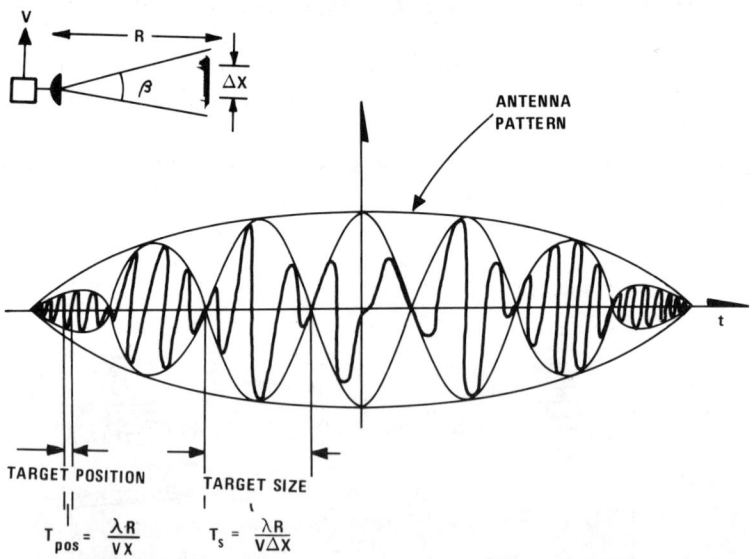

Fig. 2.22 *Period of the amplitude modulation gives information about target size. The period of the carrier gives information about target position*

Hence
$$P(\theta) \sim \int R_E(V\psi) \, e^{-j\omega\psi \frac{V}{c}} \, d(V\psi)$$
$$P(\theta) \sim W(f)$$

By measuring the power spectrum of the interferogram through which the antenna is flying, we get a direct measure of the transverse distribution of scatterers in the target at a known range since
$$P(\theta) \sim \sigma(\psi/R)$$

In summing up our findings, Fig. 2.22 is presented. Flying at speed V through the interference pattern resulting from a target consisting of two scatterers distance Δx apart at range R, a voltage versus time pattern as in Fig. 2.22 results. Finally, in Fig. 2.23 the frequency versus time pattern is shown.

The Doppler trajectory of a single target as seen from a coherent radar is given by the relationship
$$f(t) = \frac{2\pi FV}{c} \left\{ \cos \arctan \left(\frac{R}{Vt} \right) \right\}$$

where
- $f(t)$ = instantaneous frequency from one single target
- F = microwave carrier frequency
- V = aircraft velocity
- c = speed of light
- r = minimum distance to target from aircraft
- t = time referred to instant when aircraft is closest to target

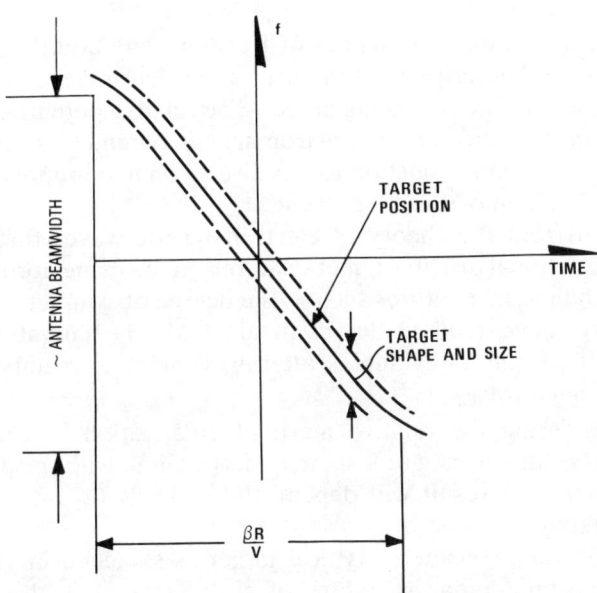

Fig. 2.23 *Freuqency/time record of a situation with one target at rest*

This relationship is calculated from the Doppler formula

$$f(t) = \frac{1}{2\pi} Vk \cos \phi$$

where

$$k = 2\pi/\lambda = \frac{2\pi F}{c}$$

ϕ = angle between V and k

2.3 Characterization of scattering centres by polarimetric methods

Up to this point we have confined ourselves to a characterization of the scattered electromagnetic field $E(K)$ in three spatial dimensions. As we have seen (see, for example, Fig. 2.16) this two-dimensional description of the scattered field gives us information about the three-dimensional distribution $\sigma(r)$ of the scattering elements constituting the scattering body. We have also in Fig. 2.16 inferred a fourth dimension, namely that of time, giving us, through Doppler considerations, information about motion.

However, there is yet one more dimension to be exploited, namely that of polarization. This property of electromagnetic waves has received considerable attention lately under the heading 'polarimetry' in particular by Boerner (1980).

Obviously, any additional degree of freedom, any additional dimension, adds to the general description of the scattered electromagnetic field and thus to the description of the scattering object. The relative potential in relation to *object recognition* of the various electromagnetic parameters deserves careful consideration. As an introduction to this discussion to be represented shortly, Figs. 2.24 and 2.25 may serve a purpose.

It is known from the theory of electromagnetic waves that polarization reveals the directional distribution of scatterers in the plane normal to the look direction of the radar. Features such as the degree of symmetry of the target, the symmetry angle (roll angle of an aircraft), the curvature and other characteristics of the individual scattering centres are embedded in the polarization signatures.

When considering the relative merit of polarization information versus multifrequency (down range) spatial information with respect to target characterization, the result will depend dramatically on the type of object being illuminated.

To illustrate this, consider a typical target as sketched in Fig. 2.24. The ultimate monostatic radar signature of such a target is the trajectory in three-dimensional space as probed by a set of two orthonormal Dirac

Matched illumination: basic principles

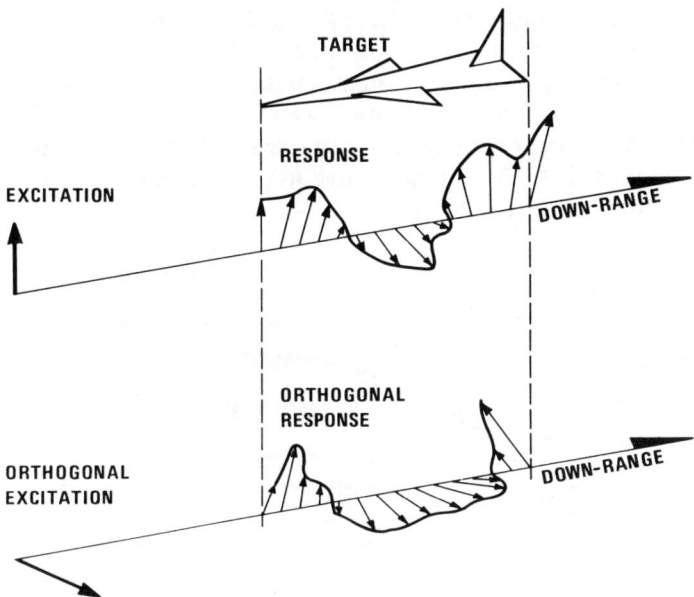

Fig. 2.24 *Ultimate signature of a target consists of two trajectories in space of electromagnetic scattered field, one for each of two orthonormal excitations (Gjessing, Hjelmstad, and Lund, 1983)*

delta-functions. At each point on the target the reflected wave will be characterized by an amplitude and direction in space for each of the two polarizations. In the case of resonance phenomena within a resolution cell on the target, a time delay will also have to be taken into account. However, for an extended target, these delayed echoes will contribute to the echoes from the following down-range cells, and will not add complexity to the target response. Note that the cross-polarized returns from each of the two orthogonal excitations are identical, owing to the symmetry of the scattering matrix for monostatic imaging through reciprocal transmission media.

For object characterization, a sufficient set of features will have to be extracted from the target response. Polarimetric information can only be used in some cases. This corresponds to no spatial information, and the down-range signature is projected into the polarization transversal plane giving the degrees of freedom of the scattering matrix.

The same number of degrees of freedom can also be obtained from multifrequency information by selecting a number of spatial resolution elements or multifrequency responses. For very simple targets, like a spheroid, the information content in the polarimetric and multifrequency domain will be the same. For more complicated real targets which require a large number of spatial resolution cells for characterization, it is to be expected that multifrequency information most readily provides adequate description of the target. This is illustrated in Fig. 2.25, which is an intuitive description of the information content in the two domains for extended targets. The information content acquired by a polarimetric radar, a multifrequency polarimetric radar and a non-polarimetric radar is plotted against spatial resolution. For the non-polarimetric multifrequency radar the span of the scattering matrix is used, thereby making the radar insensitive to the roll angle of the target. For details the reader is referred to Gjessing, Hjelmstad, and Lund (1982).

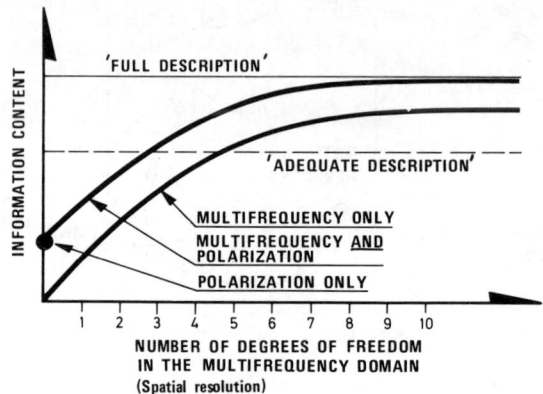

Fig. 2.25 *Illustration of the relative merit of polarization and multifrequency information (Gjessing, Hjelmstad, and Lund, 1983)*

Having briefly introduced the multi-domain adaptive radar concept also involving polarization, we shall present some very simple, and, hopefully, illustrative examples. We base these examples on a model aircraft composed of seven discrete scattering centres as illustrated in Fig. 2.26.

Here a polarization is used which excites all the scattering centres in the same manner. The delay function therefore consists of three pulses spaced $3/2\Delta z_2$ and Δz_2 as shown in Fig. 2.26.

Since the frequency covariance function $R(\Delta F) \sim V(F)V^*(F + \Delta F)$ normalized is the Fourier transform of the delay function, our radar signature will be the sum of three cosine functions with spatial period $3/2\Delta z_2$, Δz_2 and $5/2\Delta z_2$ as shown in Fig. 2.26.

We now introduce an aircraft where the scattering centres are different from the point of view of polarization. We wish to design a polarimetric multifrequency experiment from which the nature of the scattering centres can be revealed. As a basis for this discussion, Fig. 2.27 gives a survey of a set of scattering objects from the point of view of polarization characteristics.

Using, as in the experiment to be reported shortly, an adaptive multifrequency radar with full polarimetric capability, makes it possible to analyse not only the spatial distribution of the contributing scattering centres, but also the nature of these from the point of view of polarization. This implies that full polarization information is available at each microwave frequency. Two

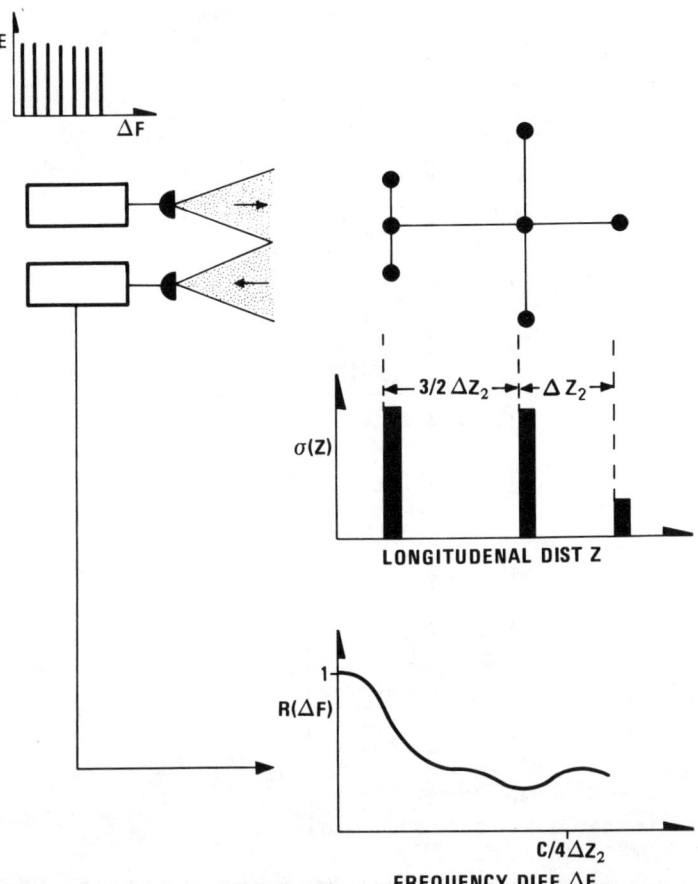

Fig. 2.26   *Multifrequency signature of an idealized aeroplane composed from seven discrete scattering centres having particular characteristics from the point of view of polarization*

orthogonal polarizations are transmitted simultaneously and on the receiving side the contributions from the co- and cross-polarized returns for each of the two receiving channels are decoded by means of a special coding technique. The scattering matrices of the radar target are thus measured and presented as a two by two complex scattering matrix (amplitude and phase).

The multifrequency signature is therefore available as five sets of signatures, each one representing one of the five independent elements of the scattering matrix. The scattering matrix can be measured by any set of orthogonal polarizations.

From a Fourier transformation point of view, a target made from a finite number of discrete scattering centres represents an ideal situation since all that is involved is the addition of N cosine functions with different amplitude and

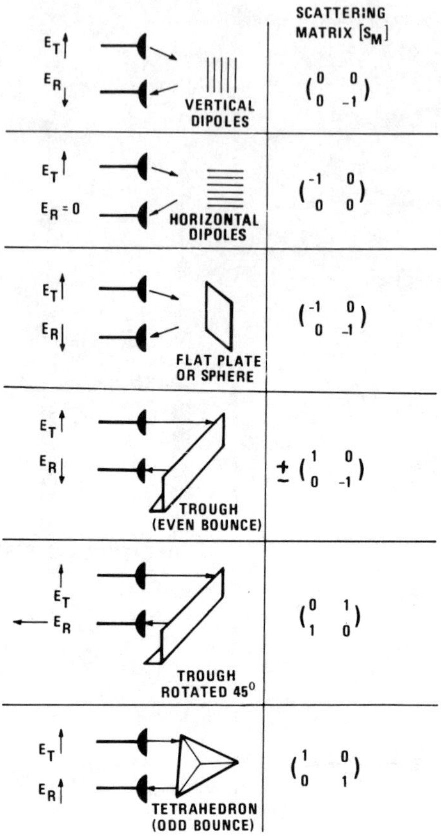

		SCATTERING MATRIX $[S_M]$
$E_T \uparrow$ $E_R \downarrow$	VERTICAL DIPOLES	$\begin{pmatrix} 0 & 0 \\ 0 & -1 \end{pmatrix}$
$E_T \uparrow$ $E_R = 0$	HORIZONTAL DIPOLES	$\begin{pmatrix} -1 & 0 \\ 0 & 0 \end{pmatrix}$
$E_T \uparrow$ $E_R \downarrow$	FLAT PLATE OR SPHERE	$\begin{pmatrix} -1 & 0 \\ 0 & -1 \end{pmatrix}$
$E_T \uparrow$ $E_R \downarrow$	TROUGH (EVEN BOUNCE)	$\pm \begin{pmatrix} 1 & 0 \\ 0 & -1 \end{pmatrix}$
$E_T \uparrow$ $\leftarrow E_R$	TROUGH ROTATED 45°	$\begin{pmatrix} 0 & 1 \\ 1 & 0 \end{pmatrix}$
$E_T \uparrow$ $E_R \uparrow$	TETRAHEDRON (ODD BOUNCE)	$\begin{pmatrix} 1 & 0 \\ 0 & 1 \end{pmatrix}$

Fig. 2.27 *Properties of certain typical scattering centres constituting a scattering object from the point of view of polarization*

period. If n is the number of scattering centres, the number N of Fourier components is

$$N = \frac{n(n-1)}{2}$$

An example of such calculations is shown in Fig. 2.28. Note that as in Fig. 2.26 the target has seven scattering centres. However, as a means of illustrating the polarization issue, we have selected four different classes of scattering centres with the following notations:

  -- scatterers for horizontal/horizontal S_{11}
  ┆ scatterers for vertical/vertical S_{22}
  | odd-bounce scatterers
  < even-bounce scatterers

Matched illumination: basic principles 41

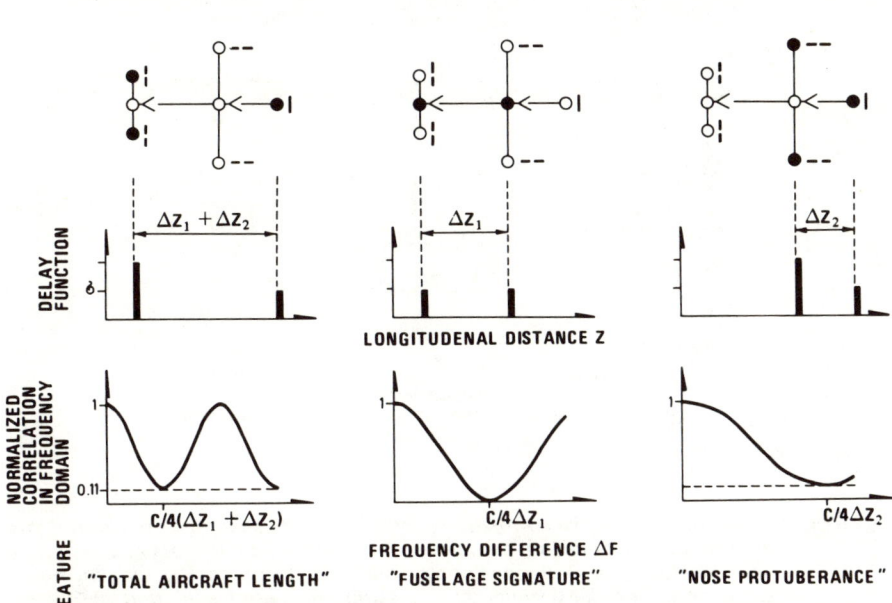

Fig. 2.28 *Measuring the frequency covariance function V(F)V*(F + ΔF) for the three appropriate polarization combinations, information can in principle be obtained about the detailed longitudinal target dimensions and about the nature of the scattering centre (Gjessing, Hjelmstad, and Lund, 1983)*

Let us leave for the present the discussion about the polarimetric multifrequency signatures with reference to Section 2.1. Now let us consider, with reference to Section 2.2, the polarimetric signatures in relation to a system with spaced antennas.

As we have already noted, the spatial correlation properties $E(X)E^*(X + \Delta X)$ of a scattered wave are obtained directly as the Fourier transform of the angular scattered power spectrum. Hence, again viewing the idealized aeroplane head on, this time with antenna elements spaced in a broadside manner, we obtain information about the relative position and nature of the scattering centres in a manner analogous to that discussed above with spaced frequency.

The results of these calculations are shown in Fig. 2.29. For details the reader is referred to Gjessing, Hjelmstad, and Lund (1983).

It is the objective of this section on polarimetry to give a brief description based on simple physics. Readers who are interested in more rigorous mathematical formulations and detailed interpretations are referred in particular to the work by Boerner (1980). Before we conclude this section, however, some detailed experimental results (Gjessing, Hjelmstad, and Lund, 1983) will be presented.

42 Matched illumination: basic principles

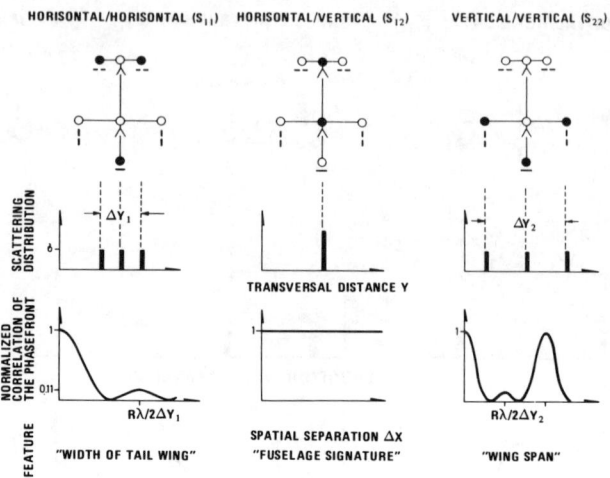

Fig. 2.29 *Measuring the spatial correlation properties of the scattered field for the three appropriate polarization combinations, information is obtained about the detailed transverse Y-dimensions of the target. Note that R is the range to the target and λ is radar wavelength (Gjessing, Hjelmstad, and Lund 1983)*

A polarimetric multifrequency radar of the author's organization has been used to investigate the signature of targets consisting of discrete polarization sensitive scattering elements. Fig. 2.30 shows the signature of a simple target with two reflectors spaced 3·75 metres apart. The lower trace is the multifrequency signature when the two reflectors are identical triple bounce reflectors, while in the upper trace, one of the reflectors is substituted by a dihedral. This is the multifrequency signature with identical transmit and receive linear polarization (the diagonal elements of the scattering matrix), and the basis is chosen so as to give zero return from the 45° oriented dihedral. Therefore, the multifrequency signature is identical to the signature of one trihedral only. With the two trihedrals, the signature is periodic, with period equal to the velocity of light divided by twice the separation of the reflectors. The fact that the correlation reaches zero indicates that the reflectors give the same echo strength.

A slightly more complicated target is shown in Fig. 2.31. Here, the two rear scattering elements (A) are 45° oriented dihedrals. Two co-located trihedrals are separated by 5 m, followed by a single trihedral 2·5 m ahead. The upper trace in Fig. 2.31 represents the diagonal elements in the scattering matrix, while the lower one represents the span of the scattering matrix (the span of the scattering matrix is the sum of all the four elements, and is equal to the total polarimetric power).

Matched illumination: basic principles

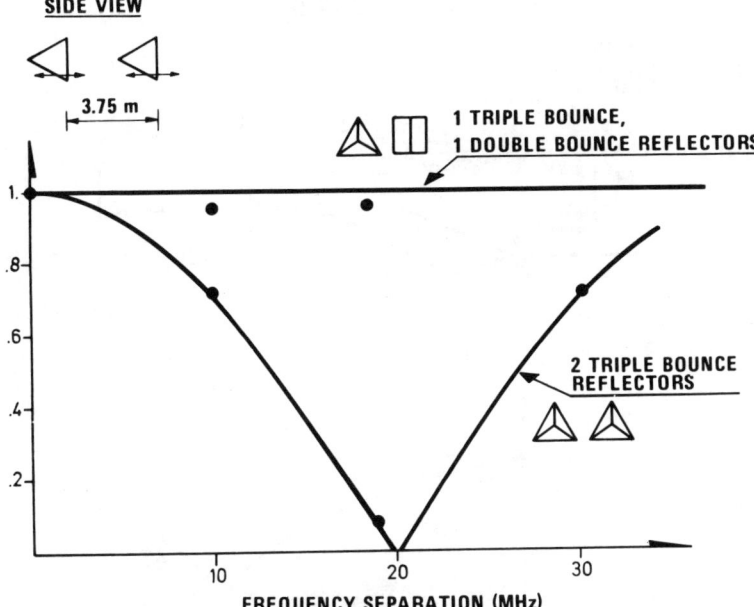

Fig. 2.30 *Polarimetric multifrequency signatures of a target consisting of two scattering elements (Gjessing, Hjelmstad, and Lund, 1983)*

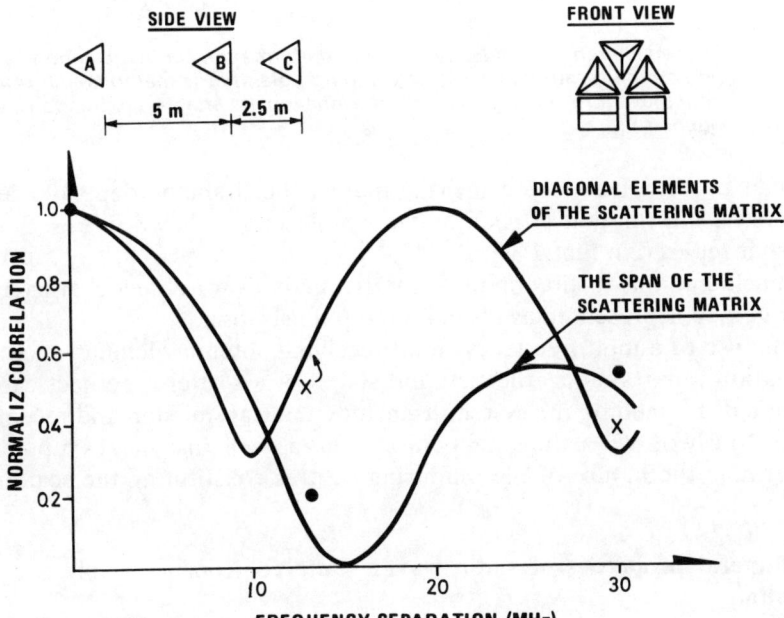

Fig. 2.31 *Polarimetric multifrequency signatures of a target consisting of five scattering elements with different polarization characteristics (Gjessing Hjelmstad, and Lund, 1983)*

44 Matched illumination: basic principles

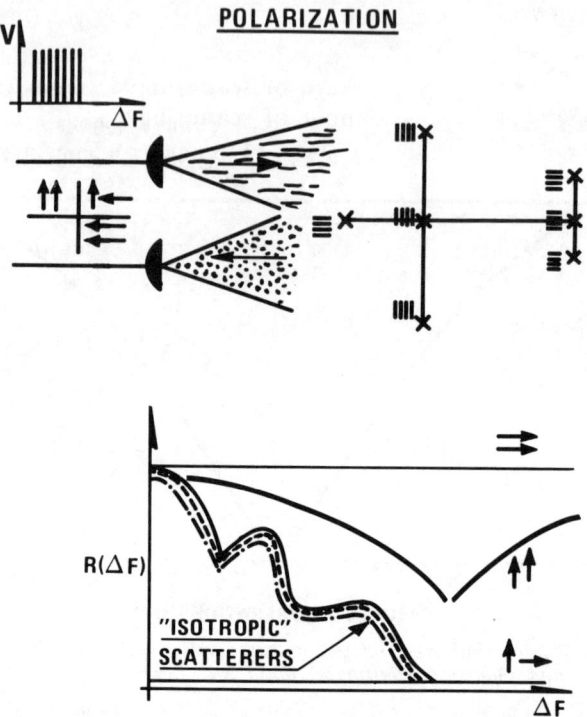

Fig. 2.32 *Matched illumination radar system gives information about the general shape and size of a scattering object. By adding polarimetric methods, information about the nature and orientation of the individual scattering centres can also be obtained*

As can be seen from these measurements, the dramatic dependence on polarimetric information as seen in the simple target shown in Fig. 2.30 is somewhat reduced in Fig. 2.31.

In concluding this section on polarimetric methods for object characterization, Fig. 2.32 is presented as a summing-up illustration.

By the use of a multifrequency, multireceiving antenna element, matched illumination radar system, the size and shape of a scattering object can be determined. Expanding the system to include the transmission and reception simultaneously of two octhogonal polarizations as well, also makes it possible to determine the nature of the scattering centres constituting the scattering object.

2.4 Temporal properties of radio waves scattered from a flexible body in motion

In introducing the concept of matched illumination, we inferred a motion pattern of the scattering surface (see, for example, Figs. 2.5 and 2.14).

Matched illumination: basic principles

We shall start this section on Doppler effects of motion by giving a general description using basic physics. We have seen that in order to extract the information of interest about shape and size of a body from a background of additive noise, signal processing methods involving temporal integration are necessary. The question then arises: how do the signals resulting from back-scattering from a surface with moving scatterers, as discussed above, decorrelate with time? This property of the signal determines the length of the integration time that we have at our disposal. In yet other applications, such as in radar study of sea state, the fluctuating return from the rough surface itself constitutes the 'signal'. Whether or not the terrain return is considered noise, or signal, it is very important to understand its temporal correlation properties and its statistics.

There are several factors that determine the temporal correlation properties of the scattered signal. These have been studied extensively over many years (see, for example, Gjessing, Jeske and Klint-Hansen, 1969). We shall present the main results. In order to ensure an understanding of the physical principles involved, we shall derive the results from first principles. Consider a scattering element within the spatial region (the scattering volume) contributing to the scattered signal. The wave incident on this has a wave vector k_1, whereas that of the scattered wave is k_s. If this scattering element is in motion relative to the observation platform – the transmitter/receiver – then the wave scattered back by the scattering element with velocity V is subjected to a Doppler shift F. This frequency shift is given by the familiar relationship

$$F = \frac{1}{2\pi}(k_1 - k_s) \cdot V$$

$$= \frac{1}{2\pi} K \cdot V \qquad (2.11)$$

Assuming a monostatic radar situation as depicted in Fig. 2.22, we measure pure back-scatter such that the angle between k_1 and k_s is 180° and the vector difference K is directed along the direction of propagation. The magnitude of

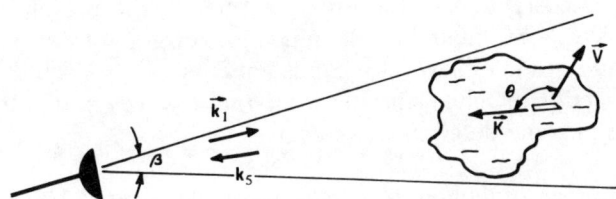

Fig. 2.33 *Scattered signal is Doppler shifted as a consequence of the motion of the scattering elements*

the vector difference K is then $4\pi/\lambda$. Thus, from one signal scattering element we have a Doppler shift F given by

$$F = \frac{1}{2\pi} \frac{4\pi}{\lambda} V \cos\theta$$

$$= \frac{2V}{\lambda} \cos\theta$$

where θ is the angle between K and V. However, we have many scattering elements contributing at any one given time. The result is that a set of waves reaches the receiver. These will have different Doppler shifts, phases and amplitudes. This situation leads to amplitude scintillations, or fading. There are two limiting cases that should be considered.

(i) *Influence on the scintillation spectrum of the velocity distribution of the scattering elements*

Let us assume that the beamwidth β is so small that the difference vector K is constant within the scattering volume. The width ΔF of the Doppler spectrum causing scintillation is then

$$\Delta F_V = \frac{1}{2\pi} K \cdot \delta V$$

$$= \frac{2\delta V}{\lambda} \cos\theta \qquad (2.12)$$

when the width of the velocity distribution is δV.

The correlation time τ is inversely proportional to ΔF. As an example, let us consider the scattered signal from a patch of the ocean surface, using a 3 cm (X-band) stationary radar. The radar senses the ripples (capillary waves) that have an oscillation frequency around 20 Hz. Decorrelation will be even faster, since each ripple oscillation only has to proceed a fraction of 2π in phase before the total back-scattered signal from all ripples has changed. In fact, a scintillation frequency of about 100 Hz is observed. Thus, every 10 ms the radar sees essentially a 'new' patch of ocean, uncorrelated to the previous one. In the presence of swell, the signal is also modulated by the much longer gravity waves, and a second correlation time of several seconds appears.

Interaction mechanisms between radio waves and ocean waves are studied in some detail in Chapter 5.

(ii) *Influence on scintillation spectrum of the spatial distribution of the scatterers; constant speed of the scattering elements*

Let us now consider the case where all the scattering elements are moving at the same velocity V. If, in addition, the difference wave vector K is also constant, there will be no Doppler broadening, but merely a constant Doppler shift causing no scintillation. Using a finite beamwidth β on our radar antenna, however, the direction of the wave vector $\mathbf{K}$ will vary within the limits of β. The corresponding Doppler broadening is then

$$\Delta F_K = \frac{1}{2\pi} \Delta \mathbf{K} \cdot \mathbf{V}$$

$$= \frac{2V}{\lambda} \{\cos\theta \cos(\theta + \beta)\}$$

which by expansion and approximation yields

$$\Delta F_K = \frac{2V}{\lambda} \sin\beta \sin\theta \qquad (2.13)$$

In the above we have not assumed any structure in the distribution of scatterers over the illuminated area. If there is such a structure, say a periodic grating with period l_0, or if we have an irregular sea surface with scale length l_0 (the width of the spatial autocorrelation function describing the irregularity structure is l_0), then this may show up in the scintillation spectrum of our radar.

Consider now two limiting cases. First assume that the beamwidth β is very narrow so that the irregularity structure of scale length l_0 is resolved. Flying with constant speed V over the irregularity structure, a fluctuation component f given by

$$2\pi f = VK$$

i.e.

$$f = \frac{V}{l_0} \qquad (2.14)$$

is obtained.

Similarly, consider the case where a large number of irregularities are illuminated by a plane wave. Let the typical scale length be l_0. As Ratcliffe originally pointed out in connection with the ionosphere (Ratcliffe, 1965), the autocorrelation function of the complex amplitude (phase included) will then have a scale length l_0 at *all distances* from the scattering region. In the same way as above, the scintillation frequency will be $f = V/l_0$, provided both phase and amplitude are measured. The result that the spatial correlation length of the complex amplitude is independent of distance may seem surprising at first. Physically, it can be explained by noting that the 'scale length information' is carried by the signals from scatterers lying far from the normal viewing direction, so that the observer is far outside the first Fresnel zone of most of the contributing scatterers.

Obviously, if our k_1–k_s vector points in the direction along which the scattering element of length l_0 is lined, the scintillation will be entirely dominated by this scale size and we get

$$f = \frac{1}{2\pi} K \cdot V = \frac{V}{l_0}$$

In conclusion then, when calculating the combined effect of all these contributions to the scintillation spectrum, several factors have to be taken into consideration. The relative importance of these depends on the geometric and geophysical conditions prevailing. For example, consider an airborne radar. Let the resolution cell on the ground be determined in azimuth angle ϕ, and at a grazing angle θ (see Fig. 2.34). This yields the following spreads in relative velocities:

$$\Delta V_{\text{azimuth}} = V \cos \theta \left(\frac{\beta}{2} \sin \theta + \frac{\beta^2}{2} \cos \theta \right)$$

$$\Delta V_{\text{radially}} = V \cos \phi \sin \theta \tan \theta \frac{c\tau}{2R} \qquad (2.15)$$

The corresponding scintillation frequency is approximately given by

$$\Delta f \simeq \frac{\Delta V_{\text{tot}}}{\lambda}, \quad \Delta V_{\text{tot}} = \{(\Delta V_{\text{az}})^2 + (\Delta V_{\text{rad}})^2\}^{\frac{1}{2}} \qquad (2.16)$$

As an exmaple, consider an X-band radar flying at 1500 m altitude at a speed of 100 m/s, with a beamwidth of 5·3° and a grazing angle of 5·6°. The azimuth Doppler decorrelation then dominates, giving a scintillation freuqency of 7 Hz straight forward, and 300 Hz in the side direction, corresponding to Doppler beats between different scatterers within the resolution cell. Depending upon the particular application, this Doppler decorrelation can be considered both as a nuisance and a blessing. On the one hand, the motion of the scatterers will mask slower time variations, making their study impossible. On the other hand, it will permit independent measurements spaced in time by $1/\Delta f$, thus making averaging easy and effective. In that case, the degree to which the target to clutter ratio can be improved by an integration process is determined by the decorrelation time of the target signal relative to that of the clutter. In our example, pointing the radar sideways gives many more independent samples than in the forward direction, assuming the signal to be the same in the two directions. These aspects will be considered in some degree of detail in Chapter 6.

Before bringing this general section on temporal properties of scattered waves to an end, a brief mention should be made of another property, namely the amplitude distribution of the signal.

We have seen that the signal received after diffuse scattering from a rough

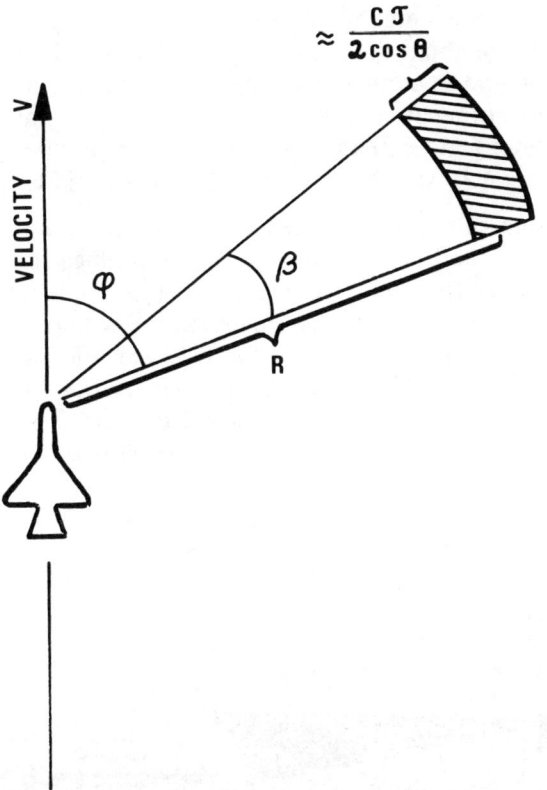

Fig. 2.34 *Notation for the Doppler width analysis*
Beamwidth is β, pulse length τ, azimuthal angle ϕ, grazing angle relative to terrain is θ (not shown)

surface is a result of many elementary waves stemming from an integral of scattering elements. There are two different multipath mechanisms that give rise to fading:

1 If the scattering elements (leaves, branches) are in motion, systematically or randomly, this motion gives rise to a Doppler frequency shift of the individual elementary waves. The power density at the receiver is thus a result of a number of electromagnetic waves that have random mutual phase, different frequencies, and, in general, different amplitudes. This situation leads to fading.

2 The scattering elements are at rest, but the refractive index integral along the ray path from the scattering element to the receiver varies. Thus the relative phase angle of the integral of elementary waves varies. This too leads to fading. As regards the amplitude distribution of the received signal, this distribution is independent of the shape of the Doppler spectrum and of its position on the frequency axes. Indeed, the amplitude distribution is

affected only by the number of incoherent scattering elements present. Thus, a knowledge about the amplitude distributions gives a rather meagre contribution to the general understanding of the structure of scattering elements contributing to the receiver field. If the number of scattering elements is large (larger than 5–6), then the amplitude distribution will closely resemble a Rayleigh distribution (Tatarsky, 1961; Ishimaru, 1978).

We shall sum up this section on general motion pattern considerations by presenting Fig. 2.35 and in so doing, we shall introduce the next section on space/time coherency (mutual coherence) giving us a measure of rigidity.

Fig. 2.35 shows that we can characterize a scattering body in three signature domains: the correlation properties in the frequency domains $V(F)\,V^*(F + \Delta F)$ give us information about target shape. The power spectrum (Doppler spectrum) associated with each ΔF (matched to scale $L = c/2\Delta F$) gives us information about the velocity distribution of the particular irregularity scale.

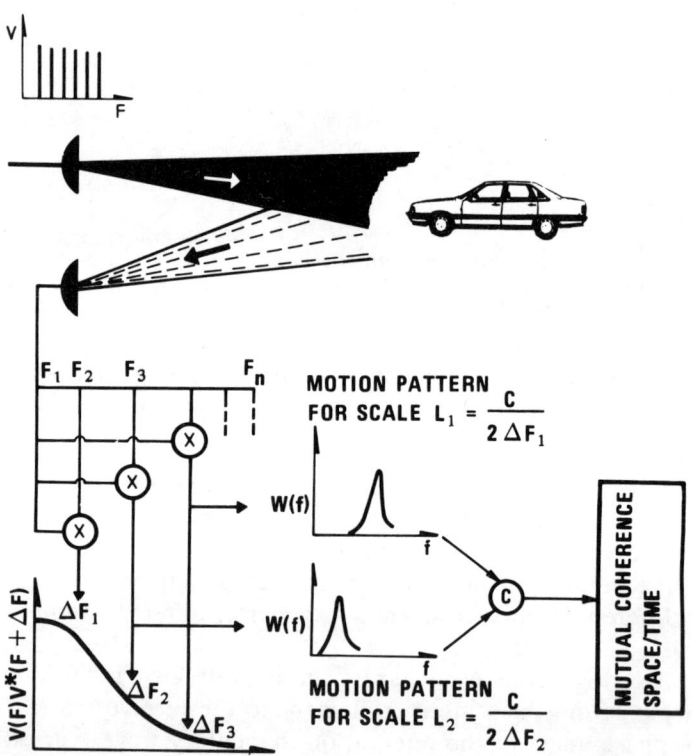

Fig. 2.35 *Scattering object can be characterized in three signature domains: shape and size, scale selective motion pattern and rigidity (space/time coherency)*

Matched illumination: basic principles 51

Finally, computing the mutual coherence, or cross-spectrum coherence, for each value opf ΔF (i.e. scale size) we obtain information about rigidity of the object, i.e. the degree to which the various scale sizes vibrate in unison.

2.5 Space/time coherence of electromagnetic waves scattered back from an object: measurement of object rigidity

We are still addressing ourselves to the problem of 'filtering out' the signature of a target in the form of, for example, a ship against a background of sea clutter. Since the sea surface exhibits a dynamic, semi-periodic irregularity structure, distributed over a large spatial region, and since the ship target manifests itself as a rigid scattering body with finite size and a non-periodic irregularity structure, the 'spatial radar signature' of this target falls into a domain which is only partly overlapping that of the sea surface. Furthermore, as will be emphasized in Chapter 5, owing to the dispersive properties of ocean waves, it is also possible, by Doppler filtering techniques, to distinguish between ship targets and contributions from the ocean background.

There is, however, still another property of the ocean surface which is yet to be explored, namely that of coherence. Illuminating a rigid body with a set of electromagnetic waves in the fashion described above, each of these radio waves is subjected to a Doppler shift which is proportional to its frequency and to the velocity of the scale size to which this wave is matched. Thus, since all the irregularity scales, constituting the total scattering cross-section of the ship target for translatory motion moves with the same velocity, the Doppler frequency shift of the various back-scattered waves will be directly related to one another. This means that if we observe the Doppler shift of the frequency pair which is coupled to, say, a 50 m irregularity scale of the ship, this will be directly correlated with the Doppler shift of the frequency pair which is coupled to the 100 m scale of the target, and the relative frequency shift will be a factor 2. This is because the scattering centres which are 100 m apart are rigidly connected to the scattering centres which are 50 m apart. Whatever changes are imposed on the velocity of the 100 m scale, the 50 m scale will follow. Referring now to Fig. 2.36, this means that if we multiply the Doppler frequency resulting from the frequency pair which is coupled to the 50 m scale size by a factor 2 and correlate this Doppler spectrum (by forming the covariance function) with that resulting from the frequency pair which is coupled to the 100 m scale size, we would expect each frequency component in the two Doppler spectra to be correlated. This is depicted in Fig. 2.36, where we have sketched the coherence function.

Now let us consider what happens to the two frequency pairs (chosen so as to couple to a scale size of 50 and 100 m, respectively) when the four waves are reflected back from the sea surface. As will be seen in Chapter 5, the 100 m wave causes a Doppler shift which is $\sqrt{2}$ times greater than that of the 50 m

Matched illumination: basic principles

wave. Furthermore, the two wave systems are not an integral part of a rigidly moving system. The 50 m wave exhibits a motion pattern which is not correlated in time or space with that of the 100 m wave.

Thus, by multiplying the Doppler spectrum resulting from the frequency pair coupled to the 50 m wave by a factor 2, and correlating this Doppler spectrum with the one obtained from the frequency pair coupled to the 100 m wave by forming the cross-spectrum coherence function, we would not expect any significant correlation. This property of ocean waves, in contrast to the corresponding properties of ship echoes, is illustrated in Fig. 2.36.

Note that, in order to optimize the effect of these coherence filtering processes, one should make use of the same number of coherence filters as there are dominating scattering centres of the target of interest. This means that in the case of the present multifrequency radar system where we are dealing with 15 frequency pairs, we could use as many as 14 'coherence filters'.

By means of such a set of coherence filters, we can determine how the coherence decreases with decreasing scale size which is inversely proportional to frequency separation ΔF. As an example, consider again the sea surface. If we correlate $V(F_1, t) V^*\{(F_1 + \Delta F), t\}$ with $V(F_2, t) V^*\{(F_2 + \Delta F), t\}$ one

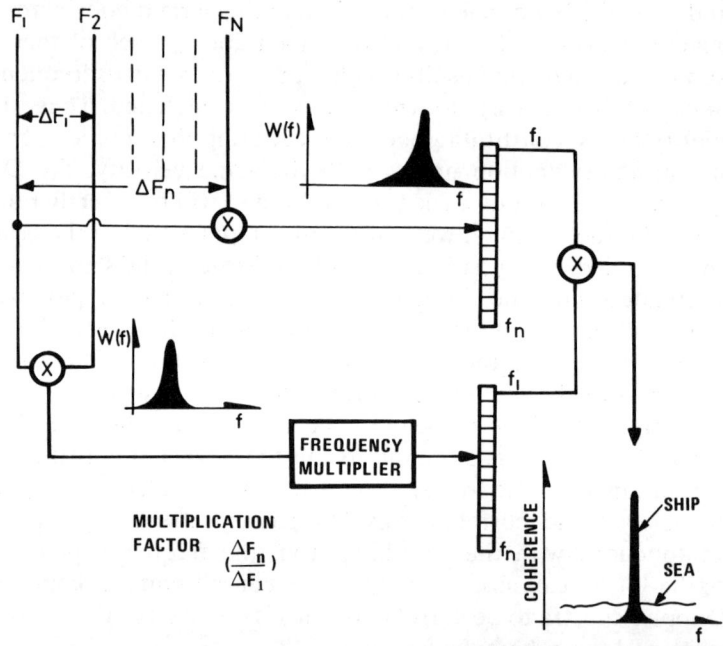

Fig. 2.36 By correlating the normalized Doppler spectrum frequency components obtained at one frequency pair with that obtained at another, thus forming the mutual coherence function, the detection and identification potential of our multifrequency radar system is considerably enhanced (Gjessing, 1981a © IEEE)

would expect a high degree of correlation since we are looking at the same ΔK at two different values of K. Integrating this wave-number correlation over time, we would expect the ratio of the coherent contribution to the incoherent one to increase since the coherent components add up in phase whereas the incoherent background (geophysical noise) gives a random phase contribution (Bass and Fuks, 1979). The larger the difference in ΔF, the smaller the covariance is expected to be, since the coupling mechanisms between ocean waves of different scale decrease with scale separation.

Thus, from the covariance curves as shown in Fig. 2.36, one for each combination of ΔF, we can plot the covariance as a function of frequency spacing ΔF. It takes little imagination to see that the same concepts can be applied to objects of substantial rigidity, such as aeroplanes.

Let us now express these coherence descriptions in mathematical terms and at the same time draw some physical conclusions. Let us express the wave-number (spatial frequency) covariance as a function of ΔK (proportional to ΔF) and time only:

$$V(F, t) V^*\{(F + \Delta F), t\} = A(\Delta F, t)$$

If we now cross-correlate $(A_1(\Delta F_1, t)$ with $A_2(\Delta F_2, t)$ we obtain information $R_{12}(\psi)$ about to what extent the intensities of the spatial scale $c/2\Delta F_1$ vary in unison with that of scale $c/2\Delta F_2$. This cross-covariance function does not, however, give us any information about the relative contribution to the correlation of the various frequency components in the two time series.

We therefore compute the Fourier transform of the cross-covariance function, thus obtaining the cross-spectrum. In the same way as the power spectrum provides information about the frequency content of a time series (the integral over all frequencies of the power spectrum in the variance), the cross-spectrum provides information in the frequency domain about the relationship between two or more time series, i.e. we obtain information about coherence and phase lag.

If the cross-covariance function is non-symmetrical, its Fourier transform is complex, thus containing even and odd terms:

$$W_{12}(\omega) \sim \int R_{12}(\psi) \, e^{-j\omega\psi} \, d\psi$$
$$= \int R_{12}(\omega) \cos \omega\psi \, d\psi + j \int R_{12}(\omega) \sin \omega\psi \, d\psi$$

The real term is referred to as the co-spectrum Co whereas the imaginary part is known as the quadrature spectrum Q.

Hence the cross-power is given as

$$|W_{12}| = \sqrt{(Co^2 + Q^2)}$$

The phase relationships (drift velocity) between the various frequencies in the two time series being cross-spectrum analysed is given by

$$\tan(\psi\omega) = \frac{Q(\omega)}{Co(\omega)}$$

Thus plotting the phase spectrum

$$\tan^{-1} \frac{Q(\omega)}{Co(\omega)}$$

as a function of frequency, information about the drift velocity is obtained directly as the slope of the curve. If we are dealing with a non-dispersive process, then the velocity is independent of wave number ΔK ($\sim \Delta F$) and we would expect a straight line relationship between phase angle θ and frequency, the slope being determined by the drift velocity. For gravity waves there is a square root relationship between phase velocity and wave number (scale) resulting in a square root relationship between θ and ω for the coherent gravity wave component. For certain regions of Doppler frequency ω, however, where the sea surface motion pattern of the various wave numbers ΔK is controlled by wind, breaking waves and other 'incoherent phenomena' rather than by coherent gravity waves, the phase relationship between the two time series $A_1(\Delta F_1, t)$ and $A_2(\Delta F_2, t)$ may be non-existent, in which case the θ versus ω curve breaks down.

It then only remains to extract one more piece of information from the cross-spectrum analysis, namely that of coherence. Since, in the general case, we are dealing with the temporal correlation between different wave numbers (spatial scales) it is appropriate to use the term mutual (space/time) coherence; this is defined as

$$Coh(\omega) = \frac{Co(\omega)^2 + Q(\omega)^2}{W_1(\omega) \ W_2(\omega)}$$

This tells us how the various frequency components in the two time series (two different spatial wave numbers) correlate.

As pointed out earlier $A(\Delta F, t)$ correlated with itself for different values of F is expected to give a high correlation for certain values of ω corresponding to the appropriate gravity wave phase velocities.

The larger the difference between the ΔKs, the poorer is the coherence expected to be, owing to the limited coupling between ocean scales. When analysing the signal scattered back from a rigid body, such as an aeroplane, however, the coherence is expected to be maintained up to small scales (large values of ΔK) since all the spatial Fourier components to which the spectrum of ΔKs are matched essentially move in unison at the same velocity, except for flutter phenomena at high values of ΔK. This important rigidity factor is of the greatest significance as a means of characterizing a scattering body.

Fig. 2.37 illustrates this target signature domain at the same time as it sums up the longitudinal one-dimensional target description.

Matched illumination: basic principles 55

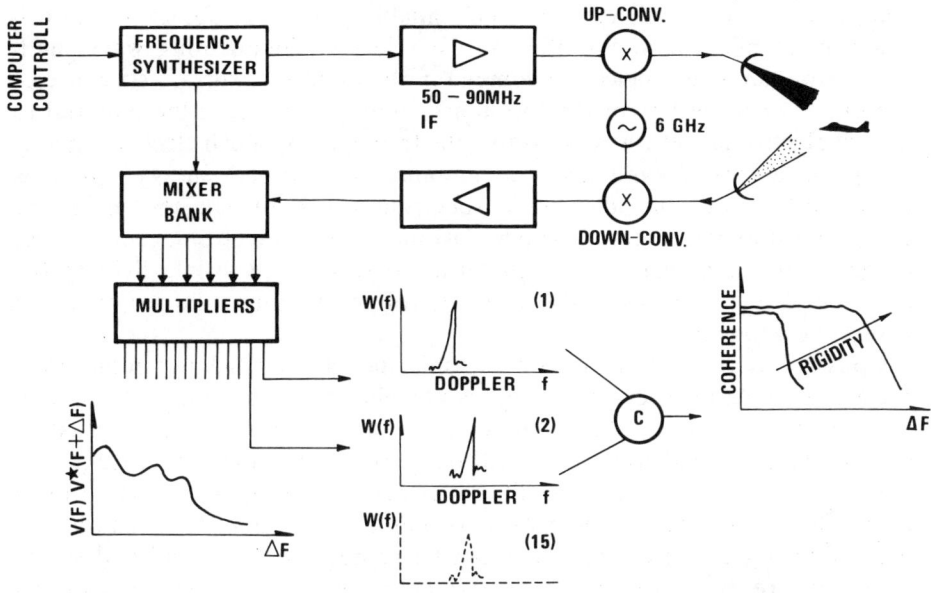

Fig. 2.37 *An object of finite rigidity can be characterized in three domains: shape and size, scale selective motion pattern, the rigidity (space/time coherence)*

2.6 Influence of absorbing surfaces: molecular resonance phenomena and illumination matched to surface material composition

Up to this point we have confined our attention to macro-scale phenomena; to resonance phenomena brought about by the interaction between electromagnetic waves and geometrical structures. These principles rely on the imaginary term in the scattering matrix; we have not considered molecular resonance phenomena giving rise to absorption.

It is not within the scope of this book to consider molecular resonance phenomena in any degree of detail. However, for the sake of completeness, and for the sake of forming the basis for a comprehensive understanding of potential mechanisms involved when electromagnetic waves are scattered from a surface, a brief highlighting of the subject may prove appropriate.

In addition to completing the scattering picture, it is believed that this section will illustrate in a rather simple way the principle of matched illumination. This is in particular the case since we are dealing with a scalar quantity rather than, as hitherto, the four-dimensional space–time problem.

First, let us dwell shortly on the fundamentals of surface spectroscopy (see, for example, Eckert *et al.*, 1974). When we discussed methods by which the geometrical shape of an object could be determined remotely, we relied on the back-scattering capability of the object. In order to determine the material composition, we shall rely on the capability of the object to absorb electromagnetic waves. Specifically, we select an illuminating wavelength which stimulates resonance phenomena in the molecular surface structure of the object. Hitting one of the resonance frequencies causes the material to absorb electromagnetic waves. Noting the frequency at which electromagnetic energy is absorbed, and also the amount of absorbed energy, gives us information about the absorption spectrum which reveals the molecular composition of the material. Having at our disposal electromagnetic waves ranging from microwaves through infra-red waves and visible light to the ultra-violet region, gives us the capability of determining a large number of material compositions.

Consider as an example a diatomic molecule within which potential forces are acting; the atoms are bound together by elastic forces as illustrated in Fig. 2.38. If these atoms, having finite mass, are excited by an alternating electromagnetic field, resonance will occur when the stimulating frequency equals that corresponding to the difference between two quantised vibrational energy levels. We can consider these two atoms, with masses m_1 and m_2, as being tied together by forces that can be represented as a ball and spring system where the restoring force obeys Hooke's law. The frequency of this simple harmonic oscillating ball and spring system is given by

$$\nu_0 = (1/2\pi)(K/\mu)^{\frac{1}{2}}$$

where K is the spring constant in Hooke's law and μ is the reduced mass, $m_1 m_2/(m_1 + m_2)$. By considering the energy which is being absorbed, we obtain

$$E_v = (v + \tfrac{1}{2})(h/2\pi)(K/\mu)^{\frac{1}{2}}$$

where $v(= 0, 1, 2, 3\ldots)$ is the vibrational quantum number.

The above example applies for two atoms. A complex molecule will have a large number of oscillating modes leading to numerous energy levels. However, if we know the detailed structure of the molecule of interest, we can in principle calculate the absorption spectrum. Also, if we have access to the material of interest, which is generally the case, then the absorption spectrum can be obtained experimentally. In both cases we have information about the 'fingerprint' and thus know what to look for. We shall now discuss methods by which chemical substances of known molecular structure (absorption spectrum) can be detected and identified selectively.

We see that, as a result of molecular resonance phenomena, the electromagnetic wave impinging on a scattering and lossy surface will suffer frequency selective absorption. These will be superimposed on the bandwidth limitations imposed by the multipath phenomena (shape and size of scattering object) discussed earlier. If one mechanism is widebanded or narrowbanded in relation to the other, we have no conceptual or practical difficulties. The difficulties arise when the width of the molecular absorption lines are

CONSIDER AS AN EXAMPLE VIBRATIONAL SPECTRA

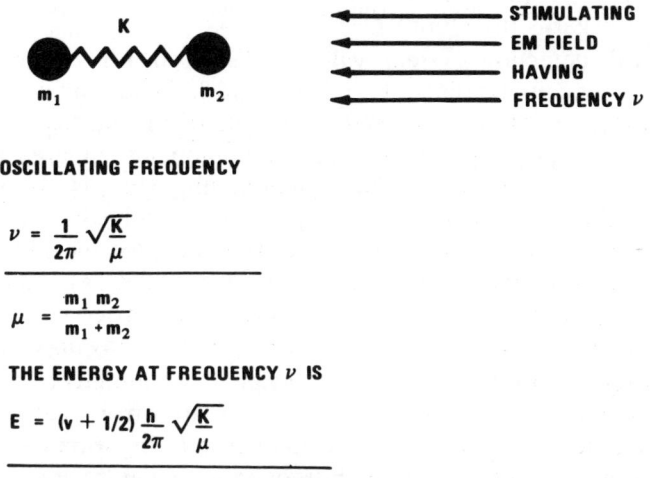

OSCILLATING FREQUENCY

$$\nu = \frac{1}{2\pi}\sqrt{\frac{K}{\mu}}$$

$$\mu = \frac{m_1 m_2}{m_1 + m_2}$$

THE ENERGY AT FREQUENCY ν IS

$$E = (v + 1/2)\frac{h}{2\pi}\sqrt{\frac{K}{\mu}}$$

$v = 0, 1, 2, 3$ etc, THE VIBRATIONAL QUANTUM NUMBER

Fig. 2.38 *An electromagnetic wave can stimulate resonance phenomena in molecules when the appropriate frequency is chosen*

comparable with the bandwidth limitations caused by the geometric shape of the scattering body ($\Delta F \sim c/2L$ where L is the size of the object).

Now, in order to illustrate the essential points and the potentials of surface spectroscopy techniques, let us define a challenging problem of practical importance, using it as an example. Our aim is to detect and identify a particular chemical agent, for example some sulphate deposited on vegetation in a certain geographical area. This could be an area contaminated by human activity or by deposits from general air pollution. *A priori* this agent may have its fingerprint (absorption lines) in any wavelength region from the ultra-violet region ($\lambda = 0.2$–$0.4\,\mu m$) through the infra-red region (2–$20\,\mu m$) to the millimetre and microwave region ($\lambda = 100$–$200\,\mu m$).

To reveal these fingerprints, we illuminate the ground on which this agent may be present with electromagnetic waves, changing the frequency over the frequency band of interest in some prescribed manner that is optimum with regard to the detection and identification of the particular chemical compound of interest. The situation is assumed to be as follows:

1 We are looking solely for a specific chemical compound.
2 We know the absorption spectrum or reflectance spectrum of this compound.
3 We know nothing about any of the other agents (interferents) that may be present.

4 We have very meagre information about the vegetation (topography) on which the agents may be deposited, but we have some general idea about the roughness of the structure (grass, coniferous trees, rocky ground). Let us assume that the background is coniferous trees.

A hypothetical detection system will be discussed for the purpose of illustrating the essential points. Fig. 2.39, which is sub-divided into two sections, shows a remote probing system in symbolic form. Section A has a generator G providing the illumination. It illuminates an area on which contaminated trees are growing. The contaminating agent is assumed to be deposited over the area, which is viewed by two receivers having two separate, not overlapping, fields of view. When the frequency of the illuminator is changed in a linear manner (saw-tooth frequency modulation) and the signals V_1 and V_2, which are received at the two receivers, are observed (referring to the symbolic diagram of Fig. 2.39), the signals have the following components:

1 A random signal with large variance (Rayleigh distributed) resulting from the topography. The signal is a result of back-scatter from scattering facets distributed in depth. They give rise to many interfering waves and result in large variations in field strength with frequency (signal trace marked 'a').
2 A deterministic signal component resulting from the complex surface chemistry (trace marked 'b') appearing as modulation of the random topography signal.
3 Deeply imbedded in the background chemistry is the component of interest, namely that stemming from the absorption spectrum of the particular chemical compound of interest (trace marked 'c').

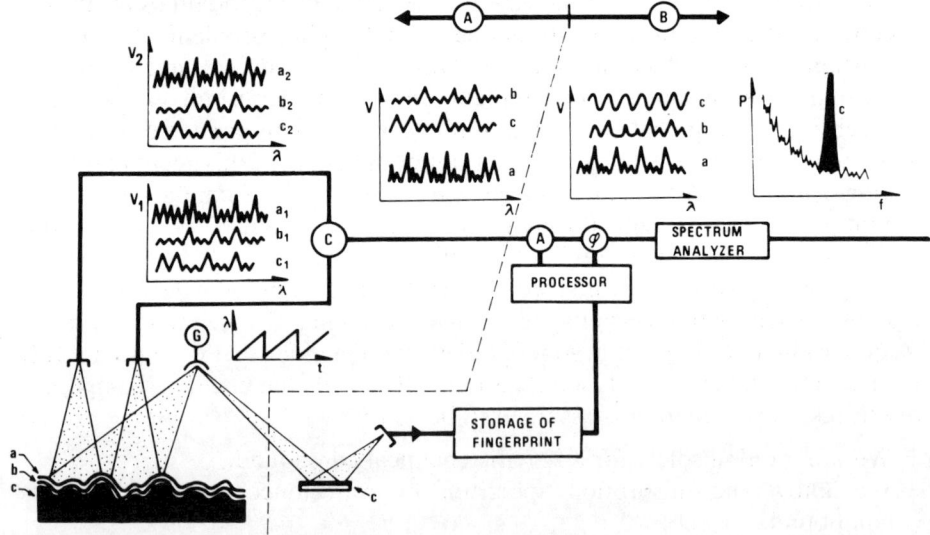

Fig. 2.39 *Symbolic diagram illustrating the optimum detection and identification principle (Gjessing, 1978a)*

Matched illumination: basic principles

Our first task in the signal enrichment process is to reduce the dominating effect of the topography (traces a_1 and a_2). In order to achieve this, we shall make use of knowledge we already have about the signal (the deterministic molecular absorption spectrum) and about the noise (the random signal resulting from scattering from a random surface such as from a tree). The noise resulting from the scattering process will obviously be statistical in nature, and there will be no correlation between the signals at the two receivers if the vegetation within the two separate fields of view is not identical in form. Furthermore, if the scatterers (the branches of the trees) move only a fraction of a wavelength from one frequency scan to the next, the correlation from one frame to another at either receiver will be limited. Since the turbulence scale (i.e. the spatial correlation distance) of the surface wind causing the vegetation to move is small compared with the height of the vegetation, the above assumptions are justified.

Therefore our task is to find a process by which the uncorrelated noise component can be reduced relative to the correlated signal component. This is done in the correlator marked C in Fig. 2.39. To ensure a thorough physical understanding of the statistical signal retrieval methods involved, this process will be discussed in some detail.

First let us consider the case where the two contributions are added together (additive noise). Fig. 2.40 gives a plot in three dimensions of the data that form the basis for the correlation process. Consider the signals from receiver number 1 first. In our Cartesian co-ordinate system the signal strength is plotted vertically, wavelength is plotted horizontally in the plane of the paper, and the frame number (or scan number) is plotted orthogonally to the $V-\lambda$ plane. The signal resulting from the deterministic absorption spectrum is denoted 'b' as in Fig. 2.39 and the random signal stemming from the vegetation is denoted 'a'. To reduce the effect of the random uncorrelated contribution, we form an ensemble covariance function for each value of λ, by comparing (correlating) $V(\lambda_1)$ from each scan. Thus we will obtain a number of points on

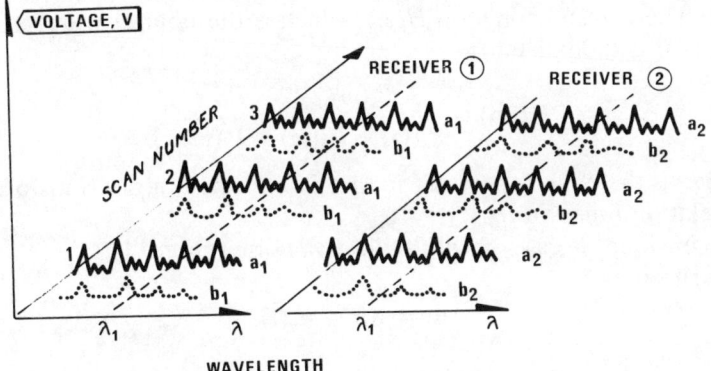

Fig. 2.40 *Data forming the basis for signal enhancement by a correlation process*

the covariance function determined by the number of frames (scans) and we will obtain a number of covariance functions determined by the wavelength resolution $\Delta\lambda$. If our ensemble then is N, our signal/noise ratio is enhanced by a factor $\sqrt{N}$.

However, the simple example with additive noise (addition of coherent and non-coherent components of the signal) is probably not the most realistic one. For example, if the coherent signal appears as an amplitude modulation on the incoherent signal, then it would be more meaningful to cross-correlate the spectral components of the signals from the two receivers by forming the coherence function instead of the covariance in the C processor.

In order to express this mathematically, we shall use an approach based on work by Kjelaas, Nordahl and Bjerkestrand (1977). In the general case we have both multiplicative and additive noise. If the spectral components of the noise are not overlapping those of the signal, a simple filtering technique can be employed. If the spectral components do overlap, we shall have to use methods that give us an optimum signal/noise ratio. First let us consider the case of *additive noise*.

If then $S(t)$ is the signal and $n(t)$ the noise, the resultant

$$F(t) = S(t) + n(t)$$

We then assume that there is no correlation between $S(t)$ and $n(t)$. The cross-covariance R_{fs} between the resultant signal f and the signal s is given by

$$R_{fs} = R_{(s+n)s}(t) = R_s(t) + R_{ns}(t) = R_s(t)$$

Similarly, the autocovariance is

$$R_f(t) = R_{s+n}(t) = R_s(t) + R_n(t)$$

This can be written in the form

$$[R_s(t) + R_n(t)] * h(t) = R_f(t)$$

where the symbol $*$ denotes convolution and $h(t)$ is therefore the impulse response. The transfer function $H(\omega)$, which is the Fourier transform of the $h(t)$ function, is then given by

$$H(\omega) = \frac{P_s(\omega)}{P_s(\omega) + P_n(\omega)} = \frac{P_s(\omega)}{P_f(\omega)}$$

where $P(\omega)$ is the power spectrum so that $P(\omega)$ is the Fourier transform of the autocorrelation function $R(t)$.

Let us consider the case of a multiplicative noise source. The resultant signal may be written as

$$f(t = S(t) * h_n(t)$$

and

$$R_f(t) * h_n(t) = R_{fs}(t)$$

which in terms of the transfer function gives

$$H(\omega) = \frac{R_{fs}(\omega)}{R_f(\omega)}$$

We can now express the transfer function $H(\omega)$ in terms of the coherence

$$coh_{sf}(\omega) = \frac{P_{sf}(\omega)}{P_s(\omega) P_f(\omega)}$$

where $P_{sf}(\omega)$ is the cross-spectrum and the transfer function is given by

$$H(\omega) = \frac{P_{sf}(\omega)}{P_f(\omega)}$$

$$= \frac{coh_{sf}(\omega) P_s(\omega) P_f(\omega)}{P_f(\omega)}$$

$$= P_s(\omega) \, coh_{sf}(\omega)$$

Thus, if we are dealing with additive uncorrelated noise, the signal/noise ratio is enhanced by forming the covariance function as implied in the discussion of Fig. 2.39. The operator C in Fig. 2.39 is a correlator giving us the ensemble covariance function. If, however, the noise is multiplicative, we shall have to cross-correlate the spectral components by forming the coherence function.

Concluding the discussion of section A in the symbolic Fig. 2.39, we note that the uncorrelated effects of the rough scattering surface are suppressed during the correlation process. All the correlated factors, the chemical compound of interest, deeply imbedded in all the uninteresting molecular structures, are retained in unaltered proportions. Now let us proceed to section B of the diagram.

This section illustrates a method by which we can make use of the detailed information that we possess about the chemical compound of interest (molecular structure, reflectance spectrum) to perform a selective detection:

1 The signal preprocessed in section A is 'filtered' with the fingerprint (reflectance spectrum) of the chemical compound of interest. In the symbolic diagram of Fig. 2.39 this is achieved by shining part of the energy from the illuminator onto a reflector containing a film of the chemical agent on which we are focusing our attention. The resulting reflectance spectrum contains the fingerprints that give us the basis for the second filtering operation.
2 The operation involves a structuring of the original frequency sweep so as to limit the frequency content of the signal resulting from the chemical agent of interest to a delta function in the frequency domain (i.e. to a pure sinusoidal variation in time over the sweep period). In a practical system one would

accomplish this by sweeping the transmitter frequency (the illuminator) in a particular manner (not with a saw-tooth as depicted in the schematic diagram) and at the same time and in synchronism with the frequency sweeping, amplitude modulate the illuminator (see Figs. 2.41 and 2.42).

To illustrate this principle and its merits, let us assume that the ϕ and A operator of Fig. 2.39 is a tape recorder. The signal is played into the tape recorder in real time and played back successively. The first operator, the 'A function', involves an amplitude modulation of the signal so as to adjust the lines of the reflectance spectrum related to the agent of interest to the same level. This involves an amplitude modulation of the reflectance spectrum of interest. The second signal from the processor (the ϕ operator) adjusts the speed of the tape to obtain constant spacing between the lines in the reflectance spectrum of interest. If the interval in the wavelength domain is large between two lines, the tape should be speeded up; conversely, if the distance is small the speed should be slowed. In this manner all the characteristic spectral lines will be equally spaced when originating from the chemical compound of interest and distributed in a disordered manner for all the interferents.

In principle, adjusting both the frequency and the amplitude of the generator illuminating the target of interest in great detail, can transform the signal component originating from our particular chemical compound to a pure sine wave requiring a scan-time limited bandwidth to be detected. In this manner the signal/noise level of our system can, in principle, be brought to a

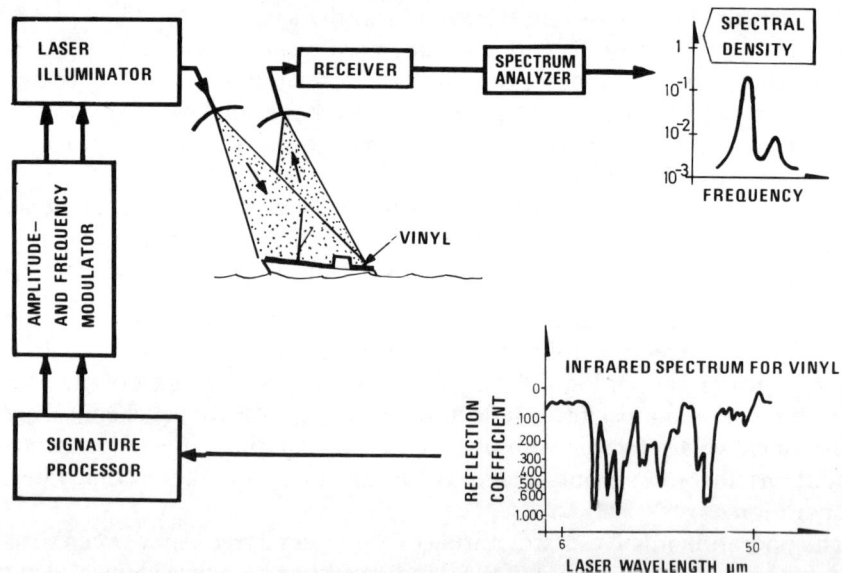

Fig. 2.41 *Illumination can be tailored to achieve optimum system sensitivity with minimum interference if detailed information about the absorption spectrum of the chemical compound of interest is available*

Matched illumination: basic principles

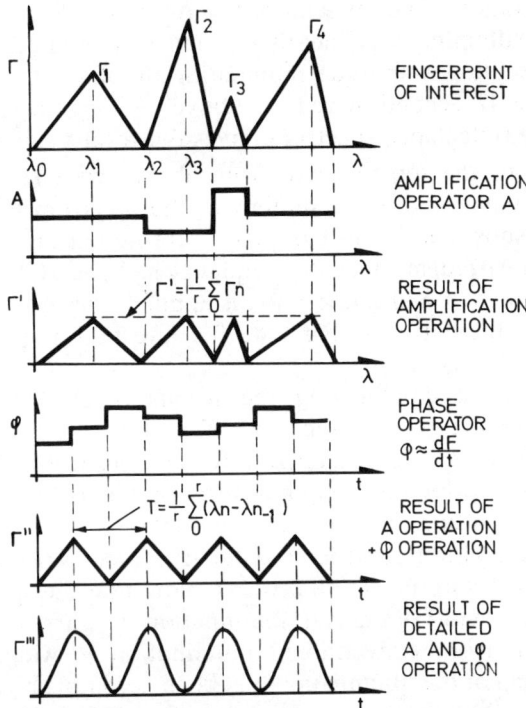

Fig. 2.42 *Example of an optimum structuring of the illumination function for the purpose of detecting/identifying a particular molecular structure (Gjessing, 1978a)*

very high level. Limiting factors are determined by the available integration time and the degree to which it is practicable to structure the illumination function.

Finally, the basic Fig. 2.39 shows the result of a final Fourier analysis of the signal. Our signal of interest is concentrated to a very narrow spectral region whereas the effect of the interferents is distributed over the spectrum. Before we proceed to consider in some greater detail the character of the noise to be suppressed, we shall give a more realistic example of a selective detection method. Such a method is illustrated schematically in Fig. 2.41.

The illumination (microwaves, IR, visible light, UV) is amplitude and frequency modulated by particular waveforms (see Fig. 2.42). These waveforms are the result of a detailed processing of the molecular signatures (see Fig. 2.41)). The processed illumination is then transmitted to the object of interest. Since the illumination is matched to the molecular structure of interest, the signal appearing at the receivers has minimum information bandwidth and the bandwidth of the entire receiving system can be minimized. Minimum bandwidth gives minimum noise contribution (Gjessing, 1978a).

Fig. 2.42 illustrates the A and φ operation process. In the top curve of Fig. 2.42 an idealized molecular spectrum is shown (spectrum of absorption, emission or reflectance). The reflectance Γ varies in a triangular manner with wavelength. If the illuminator is linearly frequency modulated, this is what the signal would look like. Amplitude modulating the illuminator in a manner illustrated in Fig. 2.42, second curve from above, seeking the same strength of all the lines in the reflectance spectrum, gives the results shown.

Referring now to the fourth curve from above, we change the relative position of the maxima and of the minima of the spectrum so as to obtain a periodic function shown in the fifth graphical representation. This is achieved by changing the rate dF/dt at which the frequency is changed. It is then obvious that if a more detailed A and φ operation is applied, the result is a sinusoidal variation, shown at the bottom of Fig. 2.42. To detect this requires only a very small bandwidth.

For the purpose of emphasizing the merits of the technique, some illustrative examples are given in Fig. 2.43. Two different absorption spectra are considered. Type 'A' is characterized by eight absorption lines, whereas type 'B' has ten absorption lines. Each absorption line is Lorentz shaped.

Adopting the amplitude and frequency modulation scheme illustrated in Fig. 2.42, the illumination is structured for Type 'A' molecules. Fig. 2.43 shows the result of this optimization process. Note that the spectrum density function associated with optimized illumination is narrow, whereas the spectrum resulting from conventional illumination is wide. Also, if an extensive structuring of the illumination had been accomplished, the resulting signal spectrum would have been a delta function. Fig. 2.43 also shows the discriminating power of the technique (right-hand figure). Here an illumination structured for type 'A' molecules is applied to structure 'B'. Note the marked difference in maximum spectral intensity and the shift in frequency of the peaks.

If a computer simulation program implementing this crude technique that involves 'first-order matching' only for three frequently encountered compounds, namely epoxy, vinyl and biphinyl, is used, the result is Fig. 2.44. Note that by the use of narrow filters as depicted in Fig. 2.41, the particular agent of interest can be selected when its absorption spectrum is known.

Let us now express this mathematically. On the basis of information about the absorption spectrum, we search the expression for the manner in which the frequency should be varied so as to give a sinusoidal variation of the received (reflected) signal. This we refer to as matched illumination.

For a linear frequency scan, the frequency as a function of time is given by

$$\omega = \omega_1 t' + \omega_0$$

where ω_1 and ω_0 are constants.

If $f(\omega)$ is the reflection/absorption spectrum, applying linear scanning we have

$$f(\omega) = f(\omega_1 t' + \omega_0)$$

Matched illumination: basic principles

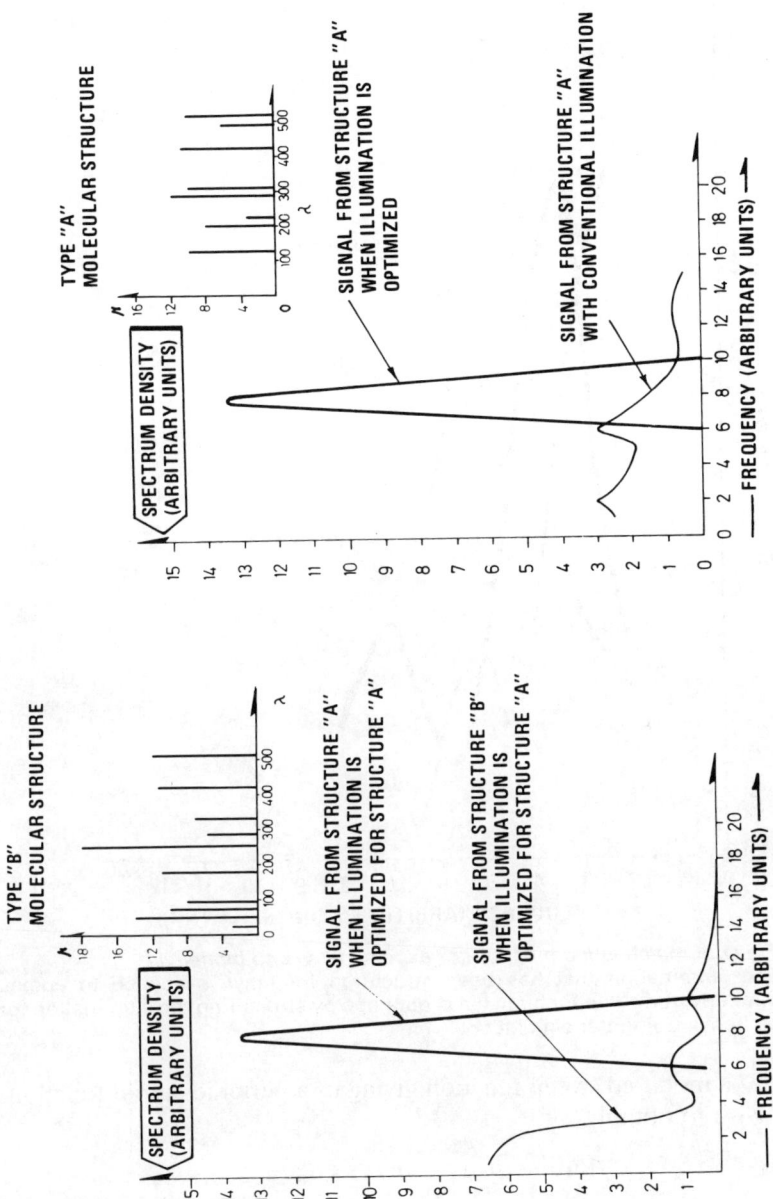

Fig. 2.43 *Effect of optimized illumination: theoretical results (Gjessing, 1978a)*

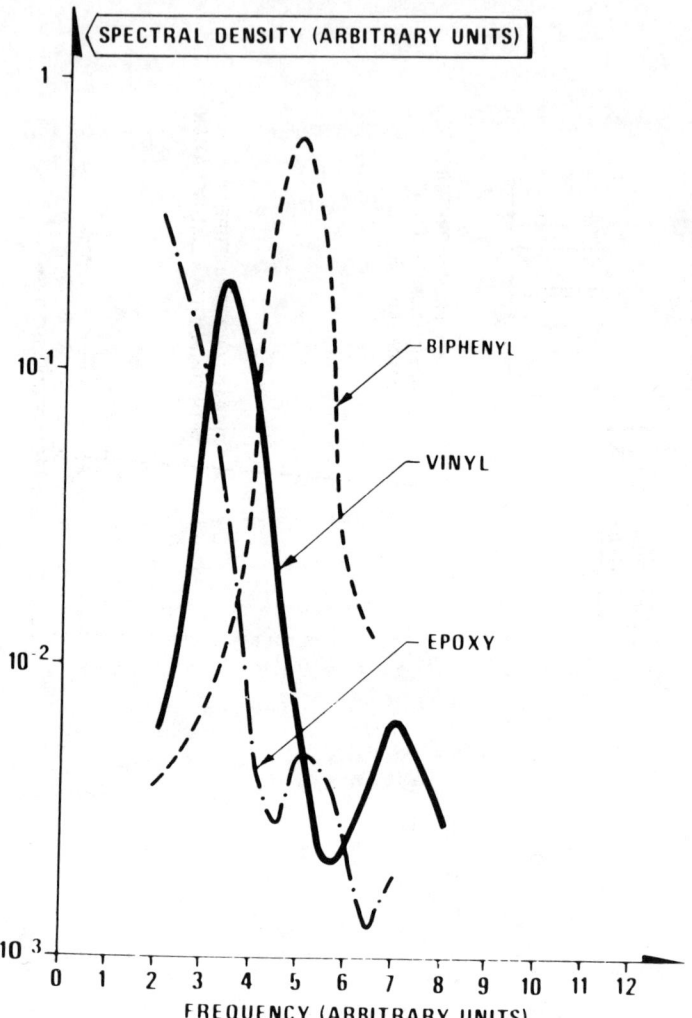

Fig. 2.44 *Effect of matched illumination for expoxy, vinyl and biphenyl*
The illumination that has been structured for vinyl is applied to epoxy. Similarly, the curve for biphenyl is obtained by structuring the illumination for biphenyl. Computer simulations.

We now want a modified sweep function giving us a periodic output function. Hence, we wish to obtain

$$f(\omega_1 t' + \omega_0) = \hat{f}(t') = g(t)$$

where $g(t)$ is our periodic function. This can be achieved by inverting the $f(t')$ function. Thus, the function should satisfy the condition

$$f^{-1}\hat{f}(t') = t' = f^{-1}[g(t)]$$

Matched illumination: basic principles

Therefore, in order to match the illumination to the agent of interest, a frequency sweep function $f^{-1}(\omega)$, which is the inverse of the reflection/absorption spectrum $f(\omega)$, should be used.

An example will clarify this. let the absorption spectrum be of an exponential form

$$f(\omega) = e^{-a\omega} = g(t)$$

$$f^{-1}[f(\omega)] \rightarrow \omega = -\frac{1}{a} \ln g(t)$$

$$\omega(t) = -\omega_1 \frac{1}{a} \ln g(t) + \omega_0$$

Having given the general expression for the matched illumination function in terms of functions describing the material of interest, we shall go on to consider the effect of the topography (the surface roughness) in relation to that of surface chemistry. The question we now ask is: how should the transmitting/receiving system be structured in order to give maximum information about the geometrical shape of an object? Or, how should the system be structured to have optimum sensitivity in relation to a given object of known size and shape?

This will be the topic of the subsequent sections.

Chapter 3
Characterization of scattering objects by means of an adaptive multifrequency radar; experimental examples

We have discussed the basic principles for matched illumination radars. We have noted that in general we are faced with the consideration of three 'filter functions': the transmission medium between the transmitter and the target, (Gjessing 1981a) the target itself, and the background against which the target is viewed.

We shall now concentrate for a moment on the target itself, and in the form of practical experimental results illustrate some of the more important points.

In order to produce a self-consistent section on solid targets, and optimum physical understanding, we shall introduce this section by summarizing briefly the essence of Chapter 2 in the context of solid scattering objects.

To provide 'matched illumination' in relation to a given target, we shall, as we have seen, have to structure the illumination both in space and in time. If we have at our disposal a radar system which can be amplitude modulated (a pulsed radar), then we shall have to shape the radar pulse in such a way as to obtain maximum influence on the returned radar pulse by the particular target of interest. If we have at our disposal a radar illuminator which can be structured in the frequency domain, then we should compose an illuminating frequency spectrum in such a way as to obtain constructive interference by all the reflecting facets of the target.

In this brief contribution we shall concentrate on the multifrequency radar system. As we shall see, this system lends itself directly to simple computer control in a manner which is very familiar to the computer scientist.

Having structured the illumination in the time domain for optimum coupling to the target, it remains to shape the phase front in space so as to obtain maximum coupling to the particular reflecting structure of interest. By making use of a matrix antenna (two-dimensional broadside array) as the radar receiver, the phase and amplitude at each receiver element can be controlled by a computer system so as to provide an antenna system which is matched to the phase front of the wave system which is reflected back from the target of interest, whereas the waves originating from the terrestrial background are suppressed. This was illustrated in Fig. 2.14.

Finally, we can manipulate the polarization properties of our transmit/receive system so as to investigate the polarization characteristics (the symmetry properties) of the target as discussed in Section 2.3.

If the four signature domains which we have at our disposal (space, frequency, polarization and motion pattern) are statistically orthogonal as regards target, propagation medium and background, we have a simple situation from an information retrieval point of view. We can treat each domain separately and the information processing system (multisensor data fusion) becomes comparatively simple. There are well established algorithms for such data handling. Examples are Kalman filtering, maximum entropy methods, optimum parameter estimation methods etc. If the degree to which the signature domains are orthogonal vary with the conditions prevailing, and if also the signatures themselves (target, background, transmission medium) vary rapidly with time, the adaption process becomes increasingly complicated.

As introduced in Chapter 2, we shall be considering four signature domains:

(a) By measuring the correlation properties $R(\Delta F)$ in the frequency domain of the waves scattered back from the illuminated area (target against background) we obtain information about the longitudinal distribution of the scatterers. Specifically it was shown that if we describe the distribution in range (longitudinally) of the scatterers by the *delay function* $f(z)$ which has dimension field strength and is the square root of the scattering cross-section $\sigma(z)$, then the correlation function in the frequency domain $R(\Delta F)$ is the Fourier transform of the autocorrelation function of $f(z)$. A measure of $R(\Delta F)$ is obtained, as we know, by multiplying the scattered field strength of frequency F by the complex conjugate of the field strength at frequency $F + \Delta F$.

Thus we have

$$E(F)\, E^*(F + \Delta F) \sim R(\Delta F) \sim FT\{\langle f(z)\rangle^2\} \tag{3.1}$$

Note that this expression refers to a rough surface, implying that the 'scattering centres' cannot be point scatterers but must be distributed over a distance in range which is comparable with the wavelength of the carrier frequency F. This condition is satisfied in most practical cases. The chances of receiving back-scattered energy by constructive specular reflection from two flat plates is indeed very small for realistic targets.

Note that if the target is illuminated with two frequencies spaced ΔF apart, then irregularities in the target with scale size $\Delta z = c/2\Delta F$ contribute to the scattered field.

(b) By measuring the spatial correlation properties of the field scattered back from the target in a plane normal to the direction of propagation, i.e. transversely, we obtain information about the transverse distribution of the scatterers. If then the x and y directions are orthogonal to the direction of propagation z (direction from radar to target), then we measure the field strength at the points x and $x + \Delta x$ in exactly the same way as above where we were dealing with different frequencies. As shown in Section 2.2

$$E(x)\, E^*(x + \Delta x) \sim R(\Delta x)$$
$$\sim FT\{\sigma(\frac{x}{R})\} \qquad (3.2)$$

where $\sigma(x)$ is the transverse distribution of the scatterers over the scattering body in the x direction and R is the distance to the target. This, of course, is the same as saying that the spatial autocorrelation of the transverse field-strength is the Fourier transform of the angular power spectrum of the scattered wave (angle of arrival spectrum).

If our receiving array consists of n unevenly spaced antenna elements measuring amplitude and phase of the reflected set of waves, then we can determine the spatial covariance function for $N = n(n-1)/2$ different values of Δx, implying that we can determine the angular power spectrum received from the scattering object for N different directions in the x–z plane.

(c) By measuring the temporal distribution of the scattered field (the power spectrum) information about the motion pattern of the target is obtained through the well-known Doppler relationship

$$f = \frac{1}{2\pi}\, \mathbf{K} \cdot \mathbf{V} \qquad (3.3)$$

where f is the Doppler frequency, $\mathbf{K}$ is the vector difference $\mathbf{k}_i - \mathbf{k}_s$ between the wave number $\mathbf{k}_i$ of the illuminating (incident) wave, and $\mathbf{k}_s$ the wave number of the scattered wave. $\mathbf{V}$ is the velocity of the scattering element.

Thus, if we are dealing with a target (such as the sea surface) composed of many scattering centres or facets which have different velocity, we obtain information about the velocity distribution of scale size Δz by illuminating the target with two frequencies with frequency difference $\Delta F = c/2\Delta z$ and by measuring the temporal variation (power spectrum) of the quantity

$$W(\omega) \sim V(F, t)\, V^*(F + \Delta F, t) \qquad (3.4)$$

Note that the velocity of the scattering element of scale size Δz is obtained from eqn. 3.3 by noting that it is the wavenumber $K = 2\pi/\Delta z$ of the difference frequency ΔF that enters into the Doppler equation in this case.

Hence the Doppler frequency is given by

$$f = \frac{2\Delta F V}{c} \cos \phi$$

(d) Combining now information about spatial distribution of scattering as obtained from the $R(\Delta F)$ function with information about velocity, we can obtain information about object rigidity by measuring the cross spectrum between

$$A_1(\Delta F_1, t) \text{ and } A_2(\Delta F_2, t)$$

where

$$A(\Delta F, t) = V(F, t) V^* \{(F + \Delta F), t\}$$

This was discussed in some detail in Section 2.5.

(e) By measuring the distribution of the scattering centres (the $\sigma(z)$, $\sigma(x)$ and $\sigma(y)$ functions) for each element of the scattering matrix from polarization measurements as discussed in Section 2.3, we obtain information about the symmetry characteristics of the scattering element, of which the target is composed.

Before we proceed to give experimental verifications of the simple mathematical models based on first principle physics, let us consider the basics of polarimetry in relation to a multifrequency radar system. Single frequency pulsed polarimetric radars have recently received considerable attention; in this presentation we shall confine ourselves to introducing the basic concepts involved in an experimental investigation, which is in progress at the author's laboratory, aiming at increasing the target identification potential of our multifrequency radar system. Note that the present radar (see Fig. 3.3) makes use of six correlated computer controlled frequency synthesizers which will give 15 different frequency spacings (can couple to 15 different target scales). The radar has two transmitters and two receivers with polarization control so as to enable us to determine the frequency covariance function

$$R(\Delta F) \sim V(F) V^*(F + \Delta F)$$

simultaneously for 15 frequency separations and for the three polarization combinations (horizontal/horizontal S_{11}, horizontal/vertical S_{12}, and vertical/vertical S_{22}). The radar can also operate with an adaptive polarization. That is, the radar optimizes its detection and identification capability by transmitting the optimal elliptic polarizations.

In order to introduce this concept, Figs. 3.1 and 3.2 are presented. Here we have 'modelled' a target in the form of an aeroplane making use of seven isotropic and polarization invariant scattering centres.

Note that to the scattering matrix elements merely are represented by their moduli. The actual system, however, measures their relative phase, thus

72 Characterization of scattering objects

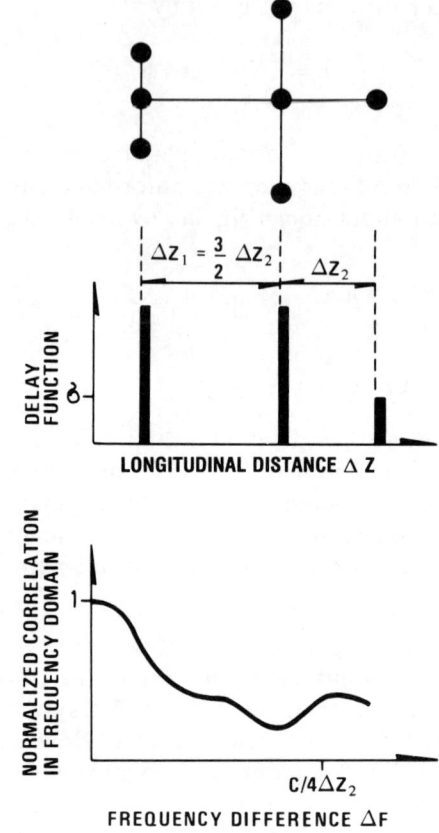

Fig. 3.1 *An idealized target consisting of isotropic polarization invariant scattering centres and the corresponding multi-frequency radar signature. The correlation function $R(\Delta F)$ is measured*

allowing detailed analysis of the target's symmetry properties within each time/space resolution cell in an arbitrary polarization basis.

Fig. 3.1 refers to the longitudinal case where the target is viewed head on with 15 coherent frequency spacings. The idealized delay function is shown and the corresponding frequency covariance function $R(\Delta F)$. In exactly the same way, Fig. 3.2 shows the spatial correlation of the scattered field when the target is viewed head on.

Note that to the extent that the target has polarization sensitive scattering centres, we can draw conclusions regarding detailed aircraft dimensions, not merely overall length, as in the case of the simplest scheme illustrated in Fig. 3.1.

Characterization of scattering objects

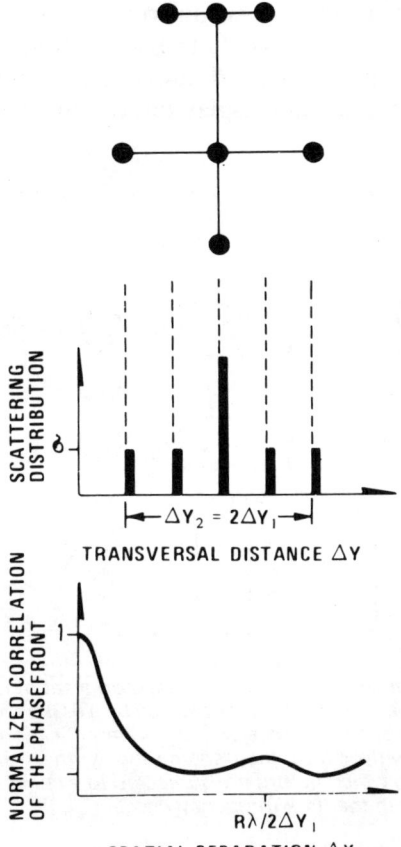

Fig. 3.2 *An idealized target viewed head on with a radar system having transversally spaced receiving antennas measuring the spatial correlation function $R(\Delta x)$*

3.1 A brief description of an experimental adaptive radar system

The radar system can, in broad terms, be subdivided into four main parts:

1. Two broadband super linear microwave transmit/receive units, one for horizontal and one for vertical polarization.
2. Six VHF computer controlled frequency synthesizers controlled by a common crystal oscillator provide for the signals to be up-converted (or down-converted upon reception of the back-scattered signal) to the 6 GHz band. The 200 mW output from the up-convertors are amplified to some 8 W and fed to a dual polarization horn antenna.
3. A second set of 6 mixers, analogue amplifiers and Doppler filters give 6 sets of terminals (in-phase and quadrature component) to be analogue-to-digital converted.

4 A powerful set of microprocessors and array processors (15 M floating point operations per sec; 32 bit words, 2 Mb RAM, 10 Mb hard disc, 50 Mb tape streamer, 0·5 Mb graphic processor) allows one to perform the appropriate computations in real time and display the various signatures characterizing the scattering object.

A schematic description of the experimental radar is shown in Fig. 3.3.

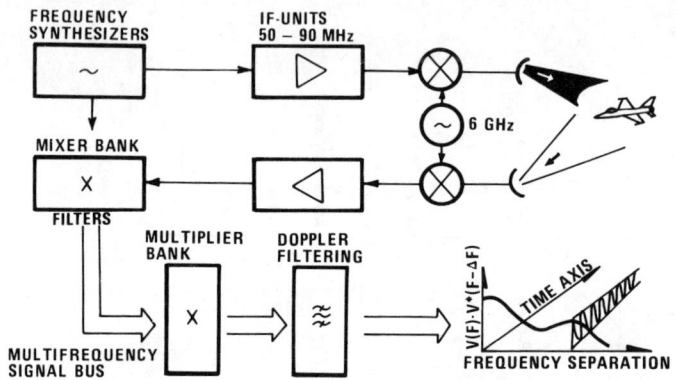

Fig. 3.3 *The multifrequency radar system is based on a set of frequency synthesizers in the 50–90 MHz band which is up-converted by a 6 GHz source which also serves at the local oscillator for the receiver. Upon amplification the receiver IF signal is mixed with the output from the transmitting synthesizers so as to obtain a set of coherent signals which in turn are subjected to a correlation analysis to give the target signature in the time/frequency domain*

Note that, since the six receiver filters are tuned separately to the same frequency as that of the continuous wave transmitter, allowing for Doppler broadening only, the system can be made very sensitive. Specifically, when the receiver bandwidth was limited to some 10 Hz (integration time 1/10 sec) a small Cessna 173 aircraft was seen with adequate signal/noise ratio to ranges in excess of 100 km with 30×30 cm antenna aperture and 12 W power transmitted per frequency line and coherent integration. The more *a priori* information one has about the target of interest (size and shape, velocity) the more narrowbanded the system can be (Gjessing, Hjelmstad, and Lund, 1982).

3.2 Ilustrative examples from experiments on rigid objects

It is the object of this section to give brief examples of simple experimental verifications so as to illustrate the main principles already discussed. Readers interested in details are referred to Gjessing, Hjelmstad, and Lund (1982).

First let us consider two extreme types of aircraft: a sleak rigid fighter plane (F-16), and a comparatively flexible, light aeroplane (Cessna 172). Tests were also performed on aeroplanes of comparable size, but different structure. These produced drastically different signatures both as regards the Doppler spectrum (flutter and vibrations superimposed on a translatory motion), and as regards the frequency covariance function (size and shape) as well as rigidity.

Fig. 3.4 shows the signature for the Cessna aeroplane. Note that, by and large, the agreement with theory when the assumption is made that the scatterers are distributed in a Gaussian manner is reasonably good. There is clear evidence, however, of a small number of scattering centres which dominate over the Gaussian distribution.

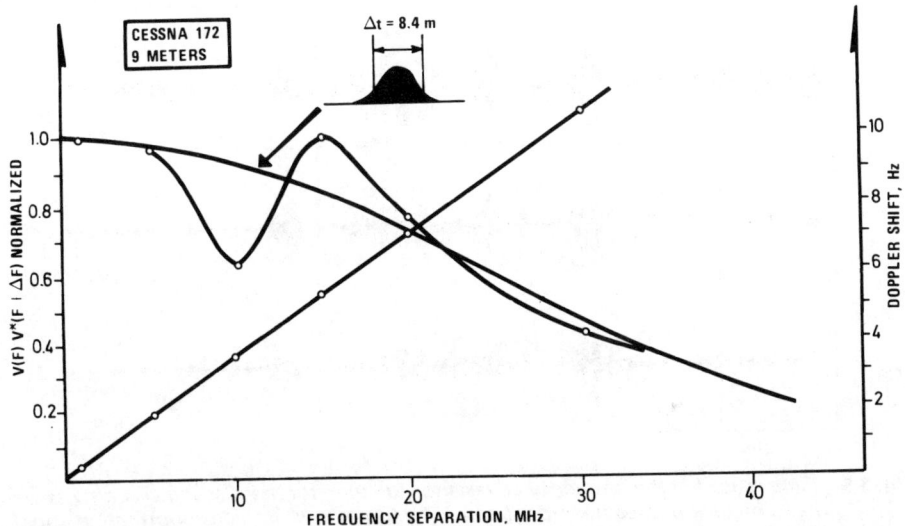

Fig. 3.4 Doppler shift and scattering centre distribution for a 9 m long Cessna 172 (Gjessing et al., 1982 © IEEE)

Fig. 3.5 shows a time record of the frequency covariance function for different values of ΔF for this aircraft. Note that the sequence of Doppler spectra suggests that we are dealing with an aircraft with a very flexible structure, as can be seen from the rapid fluctuations in the relative amplitude of the various ΔF channels.

When these records are subjected to a more detailed mutual coherence analysis many interesting features suggesting that the aircraft has a well-defined flutter component are brought out.

Another example of aircraft signatures is given in Fig. 3.6. Here the attention is focused on the very rigid F-16 aeroplane. The frequencies were selected to be matched to scattering centres distributed within the fuselage of

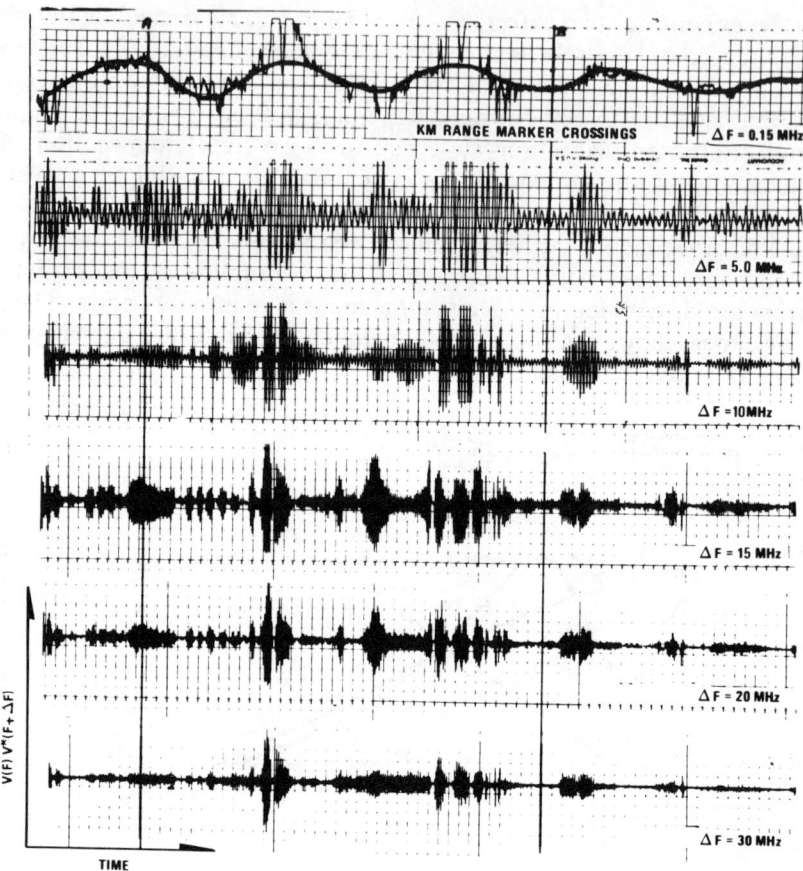

Fig. 3.5 *Time record of the frequency covariance function for the Cessna aircraft viewed head on as it passes the radar beam. The top record provides information about range marker crossings whereas the five records below give information about motion pattern for five different irregularity scales (Gjessing, Hjelmstad, and Lund, 1982 © IEEE)*

the aircraft, the longest of which is 16 m. Fig. 3.6 gives information about the longitudinal distribution of scattering centres for the F-16 aircraft. The solid line is based on theory. We here assume that the scattering centres are distributed in a Gaussian manner from nose to tail of the aircraft and that the half-power width of this delay function is 15.2 m. The experimental points are marked. Not surprisingly, the experimental results indicate that there are a few dominant scattering cross-sections in the aircraft and that smaller scatterers are distributed more or less in accordance with the Gaussian assumption. The figure also shows a plot of Doppler shift versus separation frequency. Note that these fall on a straight line in contrast to the dispersive relationship of the sea-surface gravity waves (see later).

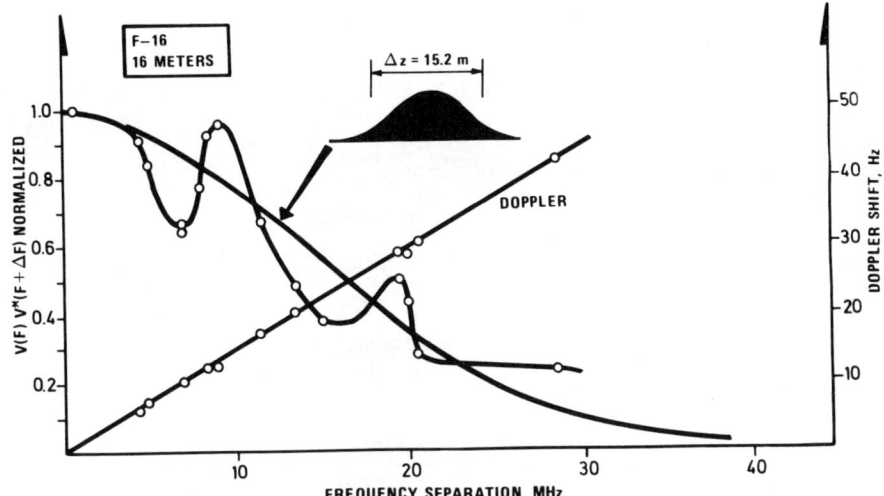

Fig. 3.6 *Longitudinal distribution of scattering centres for F-16 aircraft. Note that there are a few dominant scattering centres and apparently many smaller ones which are distributed along the main axis of the aircraft. Note also that in accordance with the theory for non-dispersive structures, there is a linear relationship between Doppler shift and illumination frequency (Gjessing, Hjelmstad, and Lund 1982 © IEEE)*

As a contrast to the Cessna aircraft (Fig. 3.5), which gives a very erratic time record, the motion pattern signature of F-16 is presented in Fig. 3.7. The trace at the top of the figure shows a 'range marker crossing'. The remaining set of time records shows that the frequency covariance varies with time for 12 different frequency separations. Note that these suggest that we are dealing with a very rigid aircraft flying on a very steady course. Note, however, that the 8 MHz time records show an amplitude modulation, the period of which is 1 s. There are many ostensible explanations for this. An obvious one is that it is due to the finite response of the autopilot. Similar modulation with less pronounced depth is found in other traces.

As a third example of rigid scattering objects, the signature of a ship (size and shape) is given in Fig. 3.8, whereas the motion pattern is shown in Fig. 3.9.

Note the very marked difference between the Doppler versus ΔF relationship of the dispersive sea surface relative to that of the rigid non-dispersive ship.

Before bringing this section on illustrative experiments to an end, the signature of the simplest of all targets, namely that of a configuration of two scattering centres, is shown.

Two small corner reflectors were spaced 15 m longitudinally apart at a distance of some 500 m from the radar. The frequency separation ΔF was varied so as to obtain constructive and destructive interference. As discussed earlier (see Fig. 2.7) the frequency covariance function $R(\Delta F)$ corresponding

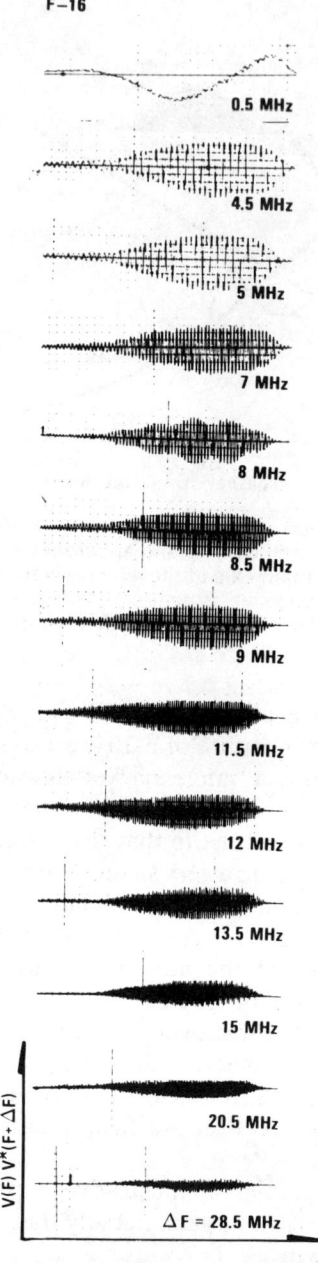

Fig. 3.7 Time records showing range marker crossing and motion pattern for the F-16 aircraft. Note the 1 Hz modulation on the 8 MHz record (Gjessing, Hjelmstad, and Lund 1982 © IEEE)

Characterization of scattering objects

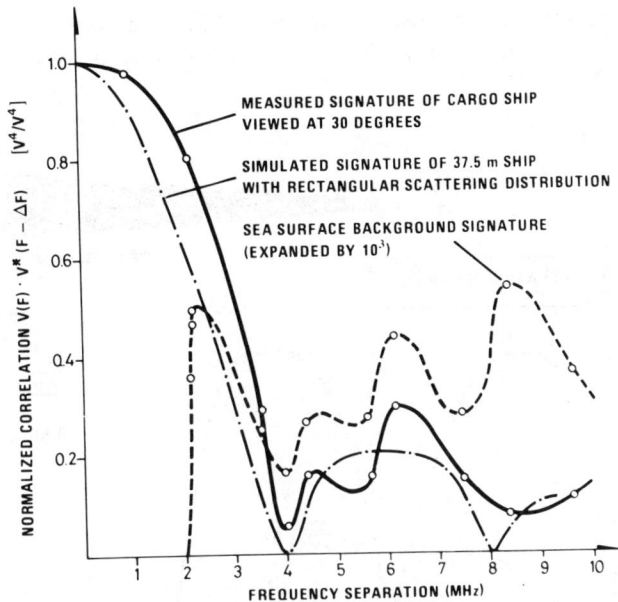

Fig. 3.8 *Normalized correlation in the frequency domain plotted as a function of frequency separation for a 37.5 m ship and for the sea surface background. Note the good agreement between measured and simulated signatures (Gjessing et al., 1983)*

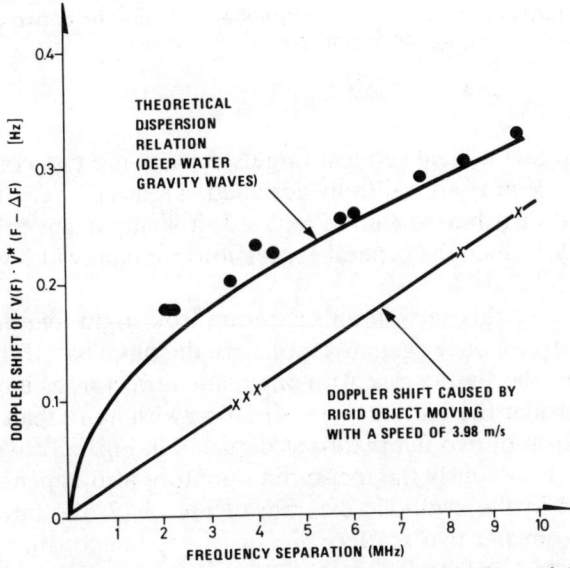

Fig. 3.9 *Doppler shift as a function of frequency separation for a rigid target and sea clutter, respectively. Note that deep sea gravity waves are dispersive: there is a square-root relationship between phase velocity and wavelength. A rigid body moving at constant velocity gives a Doppler shift that is proportional to the frequency separation and the velocity of the object*

80 Characterization of scattering objects

to a scattering object composed from two scattering centres is the Fourier transform of two delta functions. This, as we know, is a cosine function. This is illustrated in Fig. 3.10.

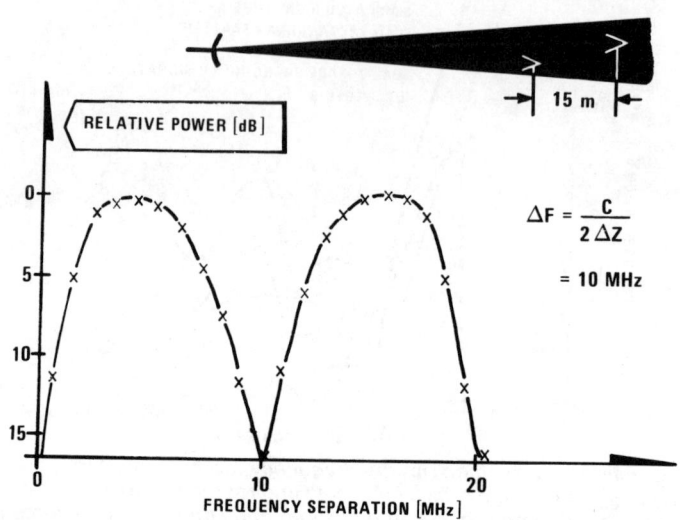

Fig. 3.10 *Signature of the simplest of all targets, a configuration of two scattering centres of equal scattering coefficient, placed 15 m apart*

When dealing with idealized test targets such as the two corner reflectors, where, if the reflectors are carefully designed, a scattering centre with little or no longitudinal distribution can be realized, it is important to remember the conditions under which the general expressions for bandwidth were made (see eqns. 2.2 to 2.4).

Before we leave this section on scattering from rigid man-made targets it may serve a purpose once again to emphasize the physics of diffuse scattering.

First consider the simple case with one plane reflector, as illustrated in Fig. 3.11. Then consider the more general situation with more than one scattering object in the form of two flat plates, as depicted in Fig. 3.12. With more than one point target obviously the measured signature also depends on K.

As indicated in the symbolic diagram of Fig. 3.12, the interaction of the contribution from the two scattering centres depends on the relative phase $e^{-jK\Delta z}$ and $e^{-j\Delta K \Delta z}$. If Δz is distributed in range over a distance comparable with $2\pi/K \approx 2.5$ cm (conditions for rough surface scattering) then the phase difference is averaged and we obtain a signature which does not depend on the exact frequency K.

Characterization of scattering objects 81

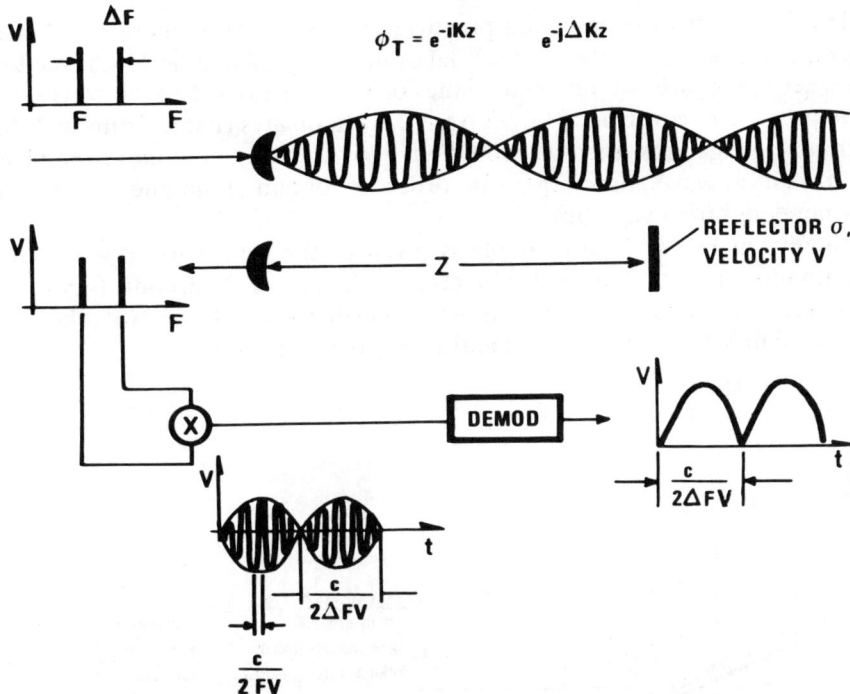

Fig. 3.11 When a target with one scattering element and zero distribution in depth moves through a two-frequency field, a simple 'fading pattern' results

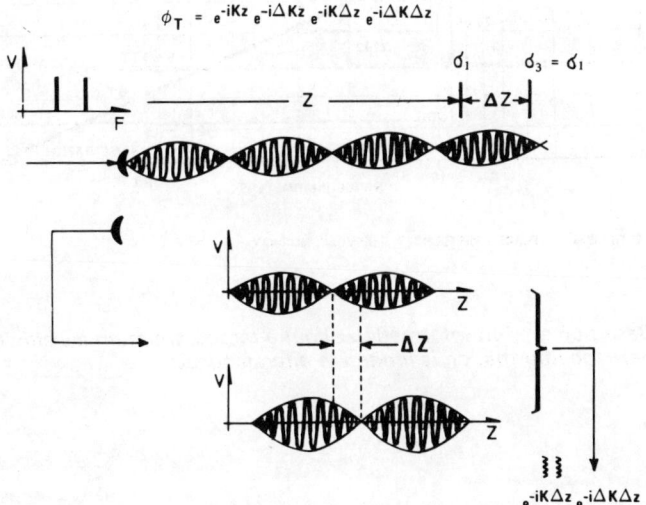

Fig. 3.12 If the scattering object consists of truly flat plates or perfect corner reflectors, then the measured signature will depend also on the carrier frequency. Under such highly unrealistic conditions the carrier frequency K will have to be frequency modulated sufficiently to change the relative phase by $\lambda/2$

This is certainly the case with practical targets, as was demonstrated in an earlier contribution by the author's laboratory (Gjessing et al 1982) (see Fig. 3.6 where there are two different values of F for the same ΔF). However, if it were possible to construct a target from truly flat plates (flat in terms of $2\pi/K$) with no corrugations, then we would have to vary K so as to move the phase front half a wavelength $2\pi/K$ in order to obtain a unique wavelength independent target signature.

Finally, in Fig 3.13 an example is given of the sensitivity of a matched illumination radar. The test object is a Cessna 172 aircraft (scattering cross-section estimated to 5 m²), flown to a distance of 100 km. Note the close relationship between theoretical and experimental results.

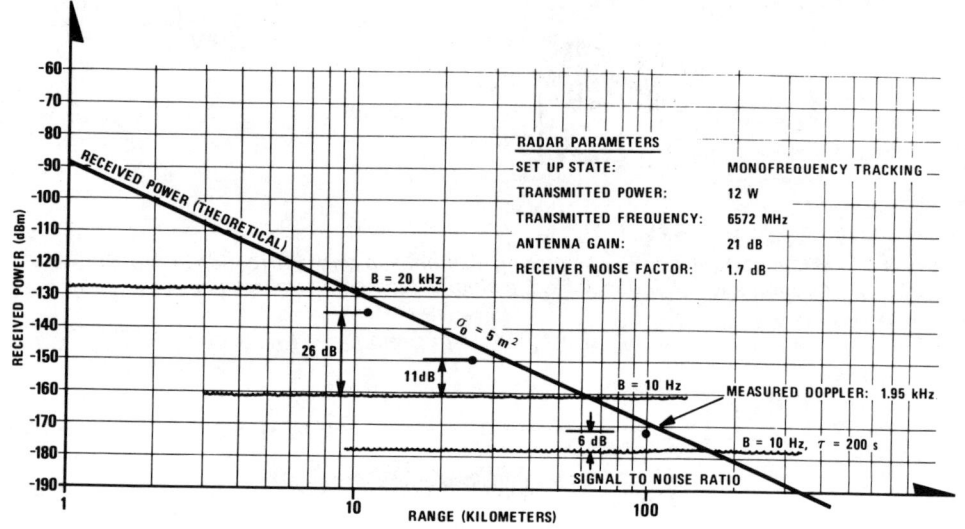

Fig. 3.13 Based on a priori information about a target, optimum sensitivity is achieved. Here the information is limited to aircraft speed

In summing up this chapter on the characterization of scattering objects, Table 3.1 may serve a purpose. Here a simple comparison is made with regard to target characterization capability between a set of modern 'high resolution' radar systems.

Table 3.1 Target characterization capability of various high resolution radar systems

		Adaptive multi-frequency	Pulse compression	Chirp (FM/CW)	SAR
Matched illumination for maximum sensitivity		yes	no	no	no
Shape/size	Depth	yes	yes?	yes	yes
	Width	no	no	no	yes
Scale selective vibration pattern		yes	no	no	no
Rigidity (space/time coherence)		yes	no	no	no
Polarization characteristics of scattering centres		yes	no	yes	?

Chapter 4

Classification of vegetation based on size, shape and motion pattern

In the previous chapter we saw that by the use of several coherent and mutually related sets of frequencies it is possible to characterize an object in four dimensions: shape and size in three dimensions if the spacial correlation properties of the received field are measured, and also scale selective motion pattern. Furthermore, we observed that by correlating the motion pattern at the various scales it is also possible to measure the rigidity of the scattering object.

Although it may be rather obvious that the same approach can be used to characterize vegetation, we shall now for the sake of completeness address ourselves to natural objects in the form of trees. Trees can be characterized, as we well know, in four domains:

1 Shape and size
2 Scale selective motion pattern
3 Mutual coherence, the degree to which the various scales constituting the object move in unison
4 Colour distribution

The question of colour has received overwhelming attention during the last decade largely stimulated by colour measuring satellites such as Landsat. As a result of a meagre spacial resolution of the existings satellites, however, the fine scale structure of the scattering objects (special wave-number spectrum, texture) have received very limited attention. During the last few years, however, we have witnessed an upsurge of interest in problems concerned with the possible effect of acid precipitation and air pollution on forest. As a result of this, the question of stress characterization of vegetation has become acute. Obviously, in order to characterize stress factors by multispectral colour measurements, one first has to identify the vegetation species. Having done this, it is conceivable that multispectral colour measurements will provide information in regard to stress. Colour measurements alone are obviously highly inadequate: when dealing with two unknowns (type of vegetation and stress) one needs a minimum of two equations, such as texture and colour. In this section, we shall confine ourselves to a discussion of shape and size together with motion pattern and limit the discussion of stress and colour by referring to the literature (see, for example, Kleman, 1985).

4.1 Characterization of vegetation on the basis of shape and size

We shall first follow the approach adopted earlier and make use of a deterministic description of the scattering object. Later on we shall use a more realistic statistical method. Obviously, when characterizing vegetation in general, and forests in particular, one is interested in average properties such as height distribution of trees, distribution of distance between branches, etc. (DeLoor et al., 1974; Eom and Fung, 1982; Ulaby et al., 1978).

Characterizing then, as before (see Section 2.1) the scattering object by its distribution in depth (range) of the scatterers, $f(z)$, the transfer function is given by

$$V\left(\frac{\omega}{c}\right) \sim E(K) \sim \int f(z) e^{-jK \cdot z} d^3z$$

Therefore, if we characterize a tree by a simple analytic expression such as an exponentially damped sinusoid

$$A \cos\left(\frac{2\pi z}{\delta z}\right) e^{-\alpha \frac{z}{\Delta z}}$$

where δz is the constant distance between branches and Δz is the 1/e height of the tree, we shall get a very simple and physically interpretable transfer function. (See Figs. 2.6 and 2.7.)

The multifrequency signature of such an idealized tree is shown in Fig. 4.1. The 1/e height is Δz, the distance between branches δz, and the distance between radar and object is z_0. Note that for the characterization method to be meaningful in practice, the distance to the object z_0 should be large in comparison with the size Δz.

To characterize a tree such as Norway spruce, which indeed resembles an exponentially damped sinusoid, a carrier frequency F which gives resonant coupling to the needle dipoles should be chosen. Since Norway spruce has needles of very uniform length (approximately 20 mm) the carrier frequency should be 7.5 GHz (to couple to Scotch pine, the carrier frequency should be approximately 3 GHz).

To couple to the uniform distance between branches (approximately 50 cm) a frequency spacing ΔF of approximately 100 MHz should be used. If then the height of the tree is typically 10 m, the 1/e width of the resonance curve depicted in Fig. 4.1 will be

$$\frac{c}{2\Delta z} = 15 \, \text{MHz}$$

86 Classification of vegetation

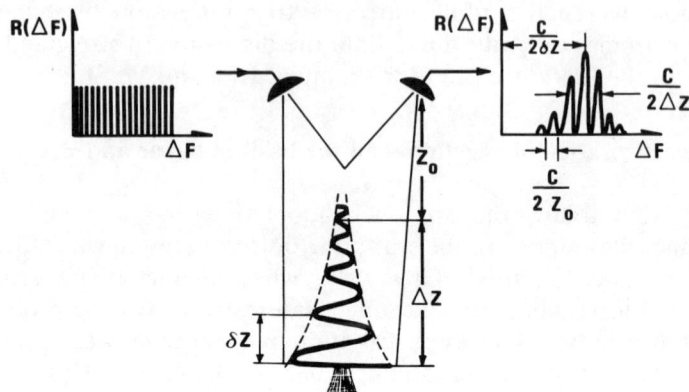

Fig. 4.1 *Multifrequency signature of a tree in the form of an exponentially damped sinusoid*

The above is for the 'deterministic' tree. In practice, obviously, even if we inspect only one specimen of Norway spruce, or any other species, there will be a certain distribution of the various characterizing dimensions.

To get some feeling for these, a typical specimen has been analysed by photographic methods as follows. A video recording of Norway spruce in a wind field was made using a CCD (charge coupled device) camera over a certain time period, which was large in comparison with the longest period of the motion pattern of the tree (see Section 4.2). By the aid of a particular texture analysis method (Gjessing, Hamran, Hjelmstad, and Aarholt, 1985) based on analogue processing of a video recording with automatic zoom capability, the three-dimensional spectrum (two dimensions in space, i.e. wave number, spectrum; one temporal dimension, i.e. power spectrum) is measured for a large wave-number region in the vertical plane of the image. As illustrated in Fig. 4.2, there are at least six different such sizes in the space domain which characterize a fir tree:

1 Height of tree
2 Distance between branches
3 Length of branches
4 Length of shoot
5 Distance between shoots
6 Length of needles

An example of such a simple and very crude analogue processing in wavenumber K-space of a specimen of Norway spruce is shown in Fig. 4.3. The numbers associated with the hatched areas (10 and 1) in the two-dimensional texture plot (K-plot) give the relative spectral intensity. Note that in addition to the height of the tree, there are three dominating scales: branch separation, shoot separation and length of needles.

Classification of vegetation 87

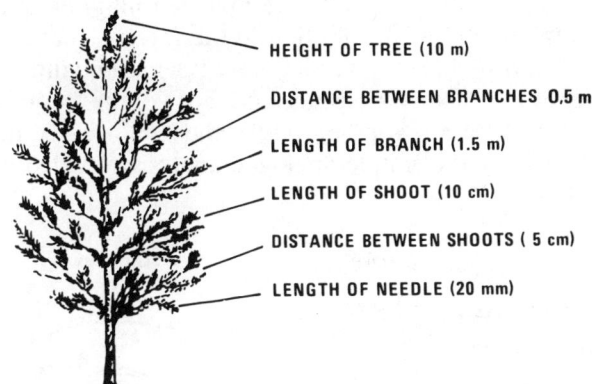

Fig. 4.2 *Several scale sizes (spectral lines in the spatial wavenumber spectrum) characterize a tree in the form of Norway spruce*

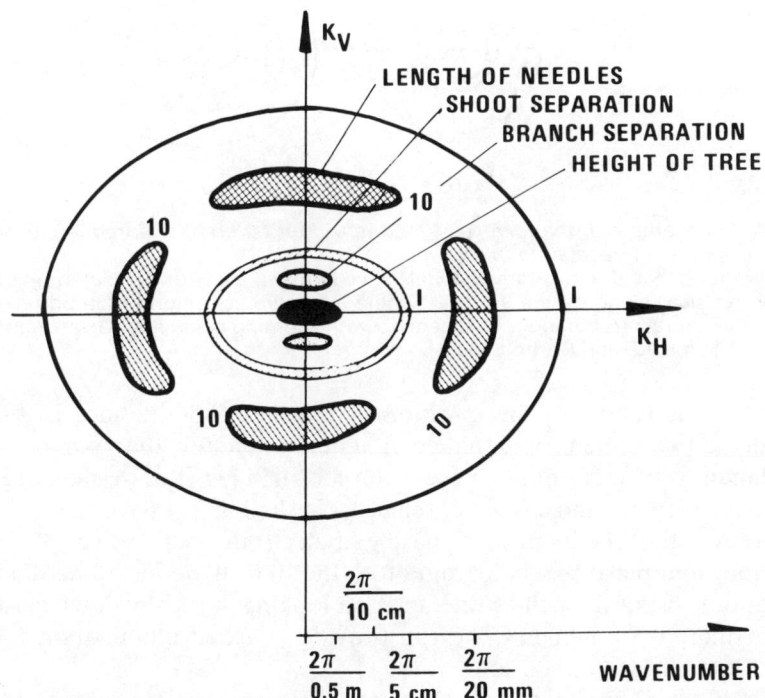

Fig. 4.3 *An example of the textural signature of Norway spruce*
Note that the two-dimensional wave-number spectrum is determined in a vertical plane by the aid of a simple analogue computing method based on a video recording and a CCD camera with zooming capability

88 Classification of vegetation

It should be emphasized that the analysis was limited to one specimen in a forest. It serves as an illustration of the matched illumination concept and should not be regarded as a contribution to the forest signature library.

It is of interest, however, to note that there is a marked difference between this textural signature of Norway spruce and those of the frequently occurring tree species, namely Scotch pine, birch and sallow. Examples of such signatures obtained by the simple analogue analysis of video recordings are shown in Fig. 4.4.

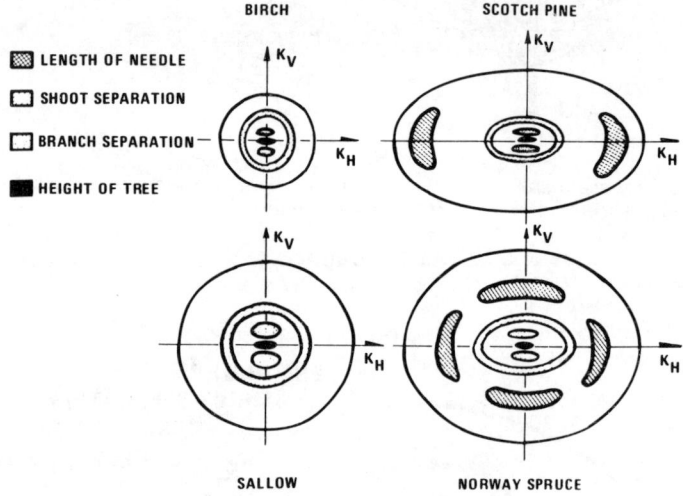

Fig. 4.4 *An example of the textural signature of four different and frequently occurring species of vegetation in Norway*
Note that the two-dimensional wave-number spectrum is determined in the vertical plane by the aid of a simple analogue computing method based on a video recording and a CCD camera with zooming capability (Gjessing, Hamran, Hjelmstad, and Aarholt 1985)

Now let us return to the question of matched illumination. In order to provide a two-dimensional matched illumination for the Norway spruce two-dimensional wave-number spectrum shown in Fig. 4.3, we shall need both spaced frequencies and spaced antennas. We shall in fact have to produce the hologram of the tree by proper phasing of electromagnetic waves of different frequency and place this hologram on to the tree. If we have one dimension only at our disposal, such as one antenna looking vertically downwards, and many coherent frequencies, we can provide matched illumination for scale sizes in the vertical dimension.

In the case of Norway spruce, therefore, the carrier frequency should be 7·5 GHz to match the 20 mm needle dipoles. There should be sidebands added to match the distance between branches ($\Delta F = 300$ MHz) and to match, for example, the length of the shoots (1·5 GHz).

We now move from the spatial dimensions, the texture, to focus our attention on the temporal dimension, on the scale selective motion pattern.

4.2 Characterization of vegetation on the basis of motion pattern

Any keen observer of phenomena in nature will have noticed the dramatic difference in the response to the wind field of the various species of trees and vegetation in general.

An example is the wave pattern characterized by a very dominating wave number set up by the wind in a wheat field. The dominating wavelength obviously provides information about mass and rigidity. These again are related to crop yield. Another example is the aspen tree. Here we note that even a low wind velocity gives rise to a very characteristic high-frequency leaf vibration which is different from any other species.

Obviously, as we have seen from earlier chapters, we can obtain information about the motion pattern associated with the various scale sizes by measuring the power spectrum (in the time domain) of each of the ΔK components scattered back from the object to which the illumination is matched. This is analogous to the investigation of the sea surface structure in space and time to be considered in the following chapter.

Furthermore, we can also, as in the case of ocean waves, determine the extent to which the various scale sizes constituting the scattering object move in unison, by computing the mutual coherence (space/time coherency) as introduced in Section 2.5.

As a means of illustrating the potential of this motion pattern analysis, simple examples will be given. Consider, as in the example given in Section 4.1, a video recording of a tree in a wind field. By simple gating techniques, the attention is concentrated simultaneously on two different points (pixels) in the dynamic image. The refreshing frequency (frame frequency) of the CCD camera is 50 Hz. Hence, each fiftieth of a second the light intensity at two different points in the image is sampled.

By Fourier analysing the output from these two gated channels we can thus, in accordance with Shannon's sampling theory, say something about the motion pattern up to a frequency of 25 Hz. Forming the cross-spectrum coherence of the output we can obviously in the same manner obtain information about the degree to which the two points in the image vibrate in unison. This question will not be addressed now.

We shall, however, give a simple example illustrating the potential of the technique by simultaneously measuring the vibration frequency (power spectrum of intensity variation from two different pixels) from a branch and a leaf, respectively, of a birch tree (Gjessing et al., 1985). To illustrate the sensitivity of the technique, the branch was cut so as to disturb the biological processes in the system.

The vibration frequency of the branch and the leaf, respectively, was measured at certain time intervals as illustrated in Fig. 4.5. As would be expected, the rigidity factor of the branch increased, owing to the loss of moisture, whereas the leaf showed a collapsing tendency.

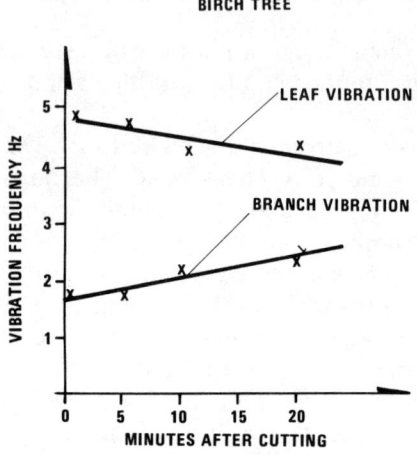

Fig. 4.5 *Vibration frequency of branch and leaf, of a birch tree in a wind field as a function of time after cutting*
Note that even during a time space as short as 20 minutes, the signature changes significantly (Gjessing et al., 1985)

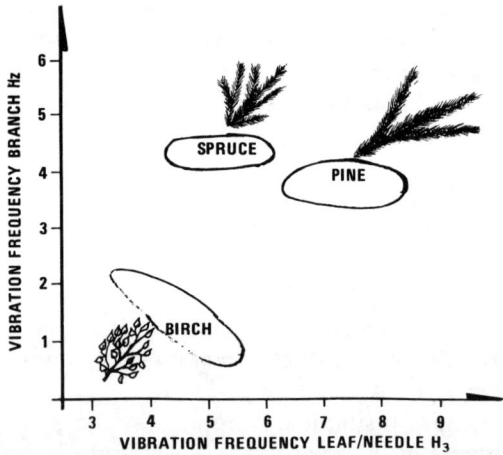

Fig. 4.6 *Preliminary results illustrating the identification potential of motion pattern analysis*
For details the reader is referred to (Gjessing, Hamran, Hjelmstad, and Aarholt, 1985)

Note that the wind field was produced by an electric fan. Note also that the tests were performed on a short branch of length some 50 cm. This may account for the marked changes even over a period as short as 20 minutes.

The experiment was repeated with a branch of spruce and a branch of pine in the same artificial wind field. Although these very preliminary results should be considered as singular sample values which may well be far from typical, they hopefully serve the purpose of illustrating the potential of motion pattern analysis.

In Fig. 4.6 the vibration frequency of the branch is plotted to the basis of leaf/needle vibration frequency.

Chapter 5

Sea clutter background: characterization of ocean surface

This section will be devoted to ocean surface phenomena. The objective is to illustrate the matched illumination principle to sea surface scattering. As before, we characterize the scattering surface by a space–time delay function $\sigma(r, t)$ which tells us how the scattering cross-section σ varies with position r and time t, or we use the Fourier transform of this delay function which is the well-known irregularity spectrum $\Phi(K, \omega)$. Note that at the moment we make no attempt to relate scattering cross-section to wave height. All we need in order to apply our adaptive radar method is information about the space–time spectrum $\Phi(K, \omega)$ of scattering cross-section.

There are many air/sea phenomena which play a role in relation to back-scattering of radio waves in the microwave region: gravity waves 'modulate' the capillary wave structure, overturning wave crests produce focusing effects and also periodic regions of strong turbulence, the boundary layer wind field with strong turbulence amplified by the ocean waves will conceivably leave a patchy and even periodic footprint on the sea surface. Internal waves originating from density gradients in the ocean further complicate the sea surface irregularity pattern.

By virtue of the fact that our particular multifrequency radar, used as an example, allows one to measure the velocity distribution ('coherent and incoherent component') associated with 15 different ocean irregularity scales simultaneously in a directional manner, it is possible to study the different air/sea mechanisms in some degree of detail.

Radio methods have a substantial potential in the study of the irregularity structure of the sea surface for several reasons:

1 They provide a rather unique possibility to measure the directional wave spectra with good directional resolution.
2 By means of a multifrequency radio method, wave length and wave velocity can be measured independently.
3 It is also possible to distinguish between coherent wave motion and incoherent motion (turbulence) and between dispersive and non-dispersive wave phenomena.

We shall present a simple theory and experimental examples of the sea surface. We shall in particular demonstrate that a microwave illuminator can be tailored so as to optimize the coupling between electromagnetic waves and ocean wave phenomena.

It is important at this early stage to specify our aims and establish perspectives. There are three rather distinctly different problem areas:

1 One must obtain an understanding of the coupling mechanisms between the electromagnetic waves on the one hand and the sea surface on the other. The scattering medium can conveniently be described as a function of space and time $f(r,t)$ or by the four-dimensional spectrum which is expressed by the Fourier transform $E(K,\omega)$ of this space–time distribution. Bandwidth measurements were introduced many years ago to determine this function characterizing the scattering medium. Particularly powerful were the methods introduced by Waterman *et al.* (1961) and Crawford *et al.* (1959). We shall be using a technique very similar to these, which was also analysed in connection with studies of waves and turbulence in the atmosphere (Gjessing 1962).

2 Having established information about the sea surface as a scattering surface for radio waves (the delay function $f(r,t)$ or the spectrum $E(K,\omega)$), it remains necessary to establish the relationship between this 'description domain' of the radio scientist and that of the ocean scientist, who wants information about the most significant wave height, wave-height spectra etc. To convert from scattering cross-section to wave height one needs information about how gravity waves and other large-scale phenomena affect small-scale phenomena (capillary waves) which are comparable with the wavelength of the radio field and which are responsible for the scattering. In this context the term 'modulation transfer function' is appropriate. Since there are many ocean surface phenomena which contribute (see Section 2.1), it is not a straightforward task to convert from scattering cross-section to wave height. Pioneers in this field of science are Wright (1968), Plant (1977), Schuler (1978), Hasselmann *et al.* (1973), Alpers and Hasselman (1978), Valenzuela (1978*a*, *b*), Barrick (1972) and others.

3 One needs an understanding of the fundamental hydrodynamic mechanisms. The list of notable scientists in this field is, of course, very long, ranging from the days of Reynolds and Richardson to present-day contributors, such as Hasselmann and Phillips.

Here we shall confine ourselves to the first problem area, the interaction between electromagnetic waves and an irregular ocean surface. We describe the sea surface in terms of the delay function $f(r,t)$ and in terms of the four-dimensional spectrum $E(K,\omega)$ and thus obtain information about the sea surface in terms of radio bandwidth. We do not enter into the very complex and often speculative field of science involved with establishing a 'modulation transfer function' to be able to obtain quantitative information about wave height. We limit ourselves to giving relative distributions. For details the reader is referred to Gjessing, Hjelmstad, and Lund (1985) and Gjessing and Hjelmstad (1986).

Before we discuss the various coupling mechanisms between electromagnetic and ocean waves, however, a brief description of the dynamic properties of the sea is of relevance.

Fig. 5.1 introduces this section. As we have already shown, a spectrum of electromagnetic waves $E(K)$ couple to the same sea surface irregularities $\Phi(K)$ (see Fig. 2.4).

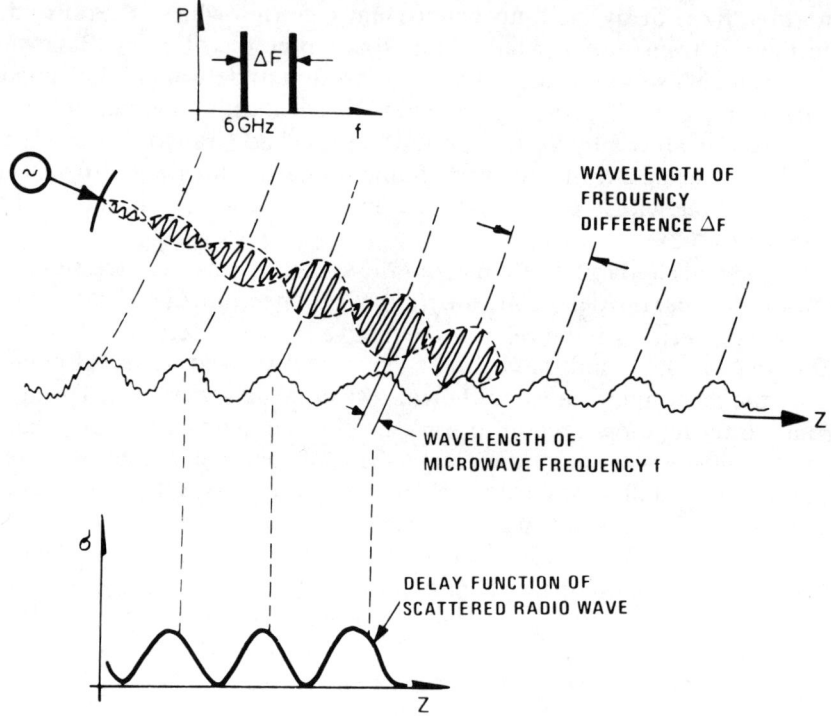

Fig. 5.1 *Electromagnetic waves E(K) interact with ocean waves when their irregularity spectrum $\Phi(K) = E^*(K)$ (Gjessing, 1981a © IEEE)*

5.1 Motion pattern of the sea surface irregularity structure

In the previous section we have shown that frequency components with mutual frequency spacing ΔF couple to irregularity scale sizes

$$\Delta L = \frac{c}{2\Delta F}$$

(To wave number $2\pi/\Delta L = 4\pi\Delta F/c$)

Any irregularity structure, periodic or random non-periodic, can be resolved into its wave-number spectrum (Fourier component) as illustrated in Fig. 2.4. Keeping track of the time history of the spatial covariance function $V(F) V^*(F + \Delta F)$, we obtain information about the motion pattern of the irregularity scale ΔL.

We shall now consider the motion pattern of the ocean surface structure. We shall briefly consider three classes of surface structures and discuss ways by which these can be separated by signal analysis methods:

1. Surface 'coherent' gravity waves resulting from a distant wind field.
2. Internal waves set up in a density gradient in the sea and coupled to the surface irregularity structures.
3. Incoherent turbulent irregularity structure resulting from a local wind field.

For the purpose of illustrating the basic principles involved, first-order expressions will form the basis for the simple calculations.

Let us first consider surface gravity waves expressed by the basic equation

$$\omega^2 = gk \tanh kd \tag{5.1}$$

where k is the wave number (spatial frequency) and d the depth.

For deep water $kd \gg 1$ and the wave equation becomes

$$\omega^2 = gk$$

The phase velocity is given by

$$v_{Ph} = \frac{\omega}{k} = \sqrt{\frac{g}{k}} \tag{5.2}$$

whereas the group velocity becomes

$$v_g = \frac{d\omega}{dk} = 1/2 \sqrt{\frac{g}{k}} \tag{5.3}$$

We have already observed that a beat frequency ΔF couples to scales $L = c/2\Delta F$. The phase velocity of gravity waves is therefore given by

$$v = \sqrt{\frac{gc}{4\pi \Delta F}} \tag{5.4}$$

Since the Doppler shift produced by a scattering element is given by

$$f = \frac{1}{2\pi} K \cdot V$$

for back-scattering this gives

$$f = \sqrt{\frac{g\Delta F}{c\pi}} \tag{5.5}$$

We note that the gravity waves are dispersive such that the Doppler shift is proportional to the square root of the illuminating freuqency ΔF matched to the irregularity scale (ocean wave length) L given by $L = c/2\Delta F$.

Then let us consider internal waves set up in an interface between water of two different densities as indicated in Fig. 5.2.

The wave equation for such waves is very much analogous with that for gravity waves on shallow water (Dysthe, 1980):

$$\omega^2 = g \frac{\delta\rho}{\rho} K \tanh kd$$

i.e.

$$v_{Ph} = \frac{\omega}{k} = \begin{cases} \sqrt{\frac{\delta\rho}{\rho} \frac{g}{k}} & \text{when } kd \gg 1 \\ \sqrt{\frac{\delta\rho}{\rho} gd} & \text{when } kd \ll 1 \end{cases} \quad (5.6)$$

For long waves on a shallow thermocline ($kd \ll 1$) we therefore have

$$v_{Ph} = v_g = \sqrt{\frac{\delta\rho}{\rho} dg} \quad (5.7)$$

i.e. no dispersion.

Note that since the density step in practice is in the order of 10^{-3} there will be practically no height modulation of the sea surface since the amplitude of the sea surface modulation is reduced by a factor $\delta\rho/\rho$ relative to that of the thermocline.

However, owing to the periodic variation of the surface current resulting from the internal waves, as illustrated in Fig. 5.2, the surface gravity waves will see a periodic current. Now let us finally briefly consider the effect of these surface currents.

Based on the principle of adiabatic invariance, it can be shown (Dysthe, 1980) that if a gravity wave is moving with a group velocity v_g in the direction of the current of velocity $\bar{u}$, the following relation applies:

$$Ev_g(v_g + \bar{u}) = \text{constant}$$

Hence, the energy E and accordingly the wave amplitude of the gravity wave must increase when the wave travels in the direction of the surface current and decrease when it travels in the opposite direction. When $v_g = \bar{u}$ the effect of the current becomes very dominating.

Referring again to Fig. 5.2, one would therefore expect certain scales on the sea surface to be particularly influenced by the internal waves, namely those which are associated with a velocity equal to that of the internal waves.

Sea clutter background: characterization of ocean surface

To obtain an expression for these scales we equate the group velocity of the surface waves and that of the internal wave. Hence

$$1/2 \sqrt{\frac{g}{k}} = \sqrt{\frac{\delta\rho}{\rho} gd} \qquad (5.8)$$

giving us a measure of the scale L on the sea surface which will be most strongly influenced by the internal wave as

$$L = \frac{2\pi}{k} = 8\pi \frac{\delta\rho}{\rho} d \qquad (5.9)$$

In order to couple to that irregularity scale with a radar system, a ΔF given by the following equation is required:

$$\Delta F = \frac{c}{2L} = \frac{c}{16\pi d} \frac{\delta\rho}{\rho} \qquad (5.10)$$

Typical values of d and $\delta\rho/\rho$ are 50 metres and 10^{-3}, respectively, giving us a coupling frequency of 119 MHz (irregularity scale 1·25 metres).

The Doppler shift caused by the internal waves is thus

$$f = \frac{1}{2\pi} K \cdot V$$

$$f = \frac{2\Delta F}{c} \sqrt{\frac{gd\,\delta\rho}{\rho}} \qquad (5.11)$$

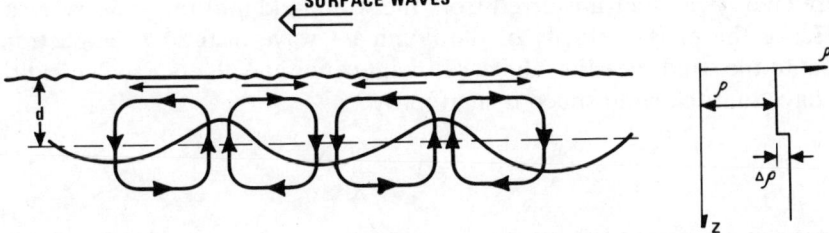

Fig. 5.2 Gravity wave will see a periodic current on the surface when an internal wave structure is present

As an illustration of interaction between internal waves and surface waves, Fig. 5.2 is presented. In contrast, note that the gravity waves give rise to a Doppler shift given by

$$f = \sqrt{\frac{g\Delta F}{c\pi}}$$

Fig. 5.3 gives experimental evidence of internal waves.

98 Sea clutter background: characterization of ocean surface

Fig. 5.3 *Internal wave pattern off the south-eastern coast of Norway depicted by SLAR Recording on 7th September 1983 by Fjellanger-Widerøe AS/Norwegian State Pollution Control Authority (Svein-Eirk Høst). The wavelength of the periodic structure is approximately 500 m. Photograph covers an area of approximately 8 by 5 km (Gjevik, 1984)*

5.2 Ocean wave spectra, wave height and wave length

Having concluded the section on motion pattern in relation to wave length, it remains to consider the wave amplitude. Wave spectra for various conditions of fetch and wind velocity have been studied by many fluid dynamicists, and in recent years in particular Phillips (1969), Pierson-Moskovitz (1964) and Hasselmann *et al.* (1973) (JONSWAP) should be mentioned. A brief sketch of the essential findings will now be given.

Phillips (1969) considered the dissipation mechanisms through breaking waves and regarded these as the limiting factor for growth. For a fully developed sea when the wind has acted on the sea for a sufficiently long time so that the phase velocity of the gravity waves is the same as the wind velocity, no more energy can be transferred from the wind field into the ocean waves.

Hence the phase velocity of the dominant wave in the wave spectrum is equal to the wind speed; we have resonant conditions. Under such conditions we have that the wind speed u is given by

$$u = v = \frac{\omega}{K}$$

where v is the phase velocity of the ocean wave, ω its angular frequency and K its wave number.

Under such conditions Phillips gives the following expression for the wave-height spectrum:

$$s(f) = \begin{cases} \dfrac{0\cdot 01 g^2}{(2\pi^4)} f^{-5} & \text{for } f > \dfrac{g}{2\pi u} \\ 0 & \text{for } f < \dfrac{g}{2\pi u} \end{cases} \quad (5.12)$$

Since the relationship between wave frequency $f = \omega/2\pi$ and ocean wave length L is given by

$$L = \frac{g}{2\pi f^2}$$

the cutoff wave length L_m is given by

$$L_m = \frac{2\pi u^2}{g}$$

The Phillips expression is shown in Fig. 5.4b. For comparison, a measured spectrum after Pierson-Moskovitz is also shown (Fig. 5.4a).

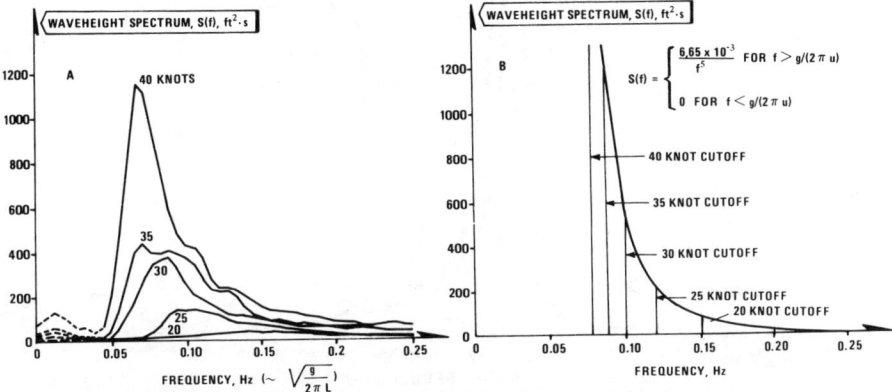

Fig. 5.4 *Measured and model wave height temporal spectra*
  (a) Measured (after Pierson-Moskovitz, 1964); seas are not fully developed at winds about 25 knots
  (b) Phillips model for fully developed seas

Pierson-Moskovitz gave a similar spectrum with the addition of an $\exp\{-(f_m/f)^4\}$ term

$$S(f) = \frac{0.01 g^2}{(2\pi)^4} f^{-5} \exp\left\{-\frac{5}{4}\left(\frac{f_m}{f}\right)^4\right\} \tag{5.13}$$

Pierson-Moskovitz measured the driving wind u at a height of 19.5 m above sea level. As before

$$f_m = \frac{g}{2\pi u}$$

Finally, the model developed by Hasselman from the JONSWAP also takes fetch into consideration. For infinite fetch, the Hasselman expression degenerates to a spectrum similar to that of Pierson-Moskovitz.

Hasselman suggests the following wave-height spectrum

$$E(f) = \frac{\alpha g^2}{(2\pi)^4} f^{-5} \exp\left\{-\frac{5}{4}\left(\frac{f}{f_m}\right)^{-4}\right\} \gamma^{\exp\{[-(f-f_m)]/(2\sigma^2 f_m^2)\}} \tag{5.14}$$

$$\sigma = \begin{cases} \sigma_a & \text{for } f < f_m \\ \sigma_b & \text{for } f > f_m \end{cases}$$

The various parameters are defined in Fig. 5.5 showing the results of the JONSWAP experiments for different values of fetch. Note that for a fully developed spectrum $\gamma = 1$ and $\alpha = 0{\cdot}01$.

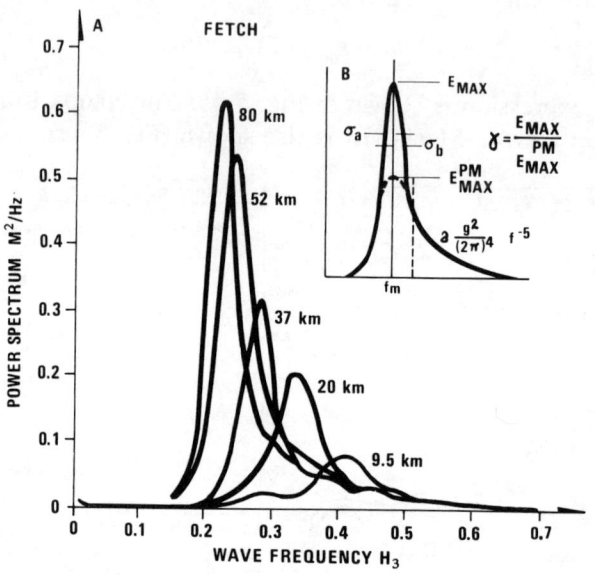

Fig. 5.5 JONSWAP experimental spectra for different values of fetch (Hasselman et al., 1973)

5.3 Interaction mechanisms between plane electromagnetic waves and sea surface irregularities: basic principles

The basic scattering equation be expressed as

$$E_s(K,t) \sim \int_{-\infty}^{\infty} f(r,t)\, e^{-jK \cdot r}\, d^3r \tag{5.15}$$

$$K = k_i - k_s \quad \text{and} \quad |K| = \frac{4\pi}{\lambda} \sin \theta/2$$

λ is wavelength and θ is the scattering angle.

In order to obtain back-scatter from a scattering surface, the surface must have a structure, regular or irregular, containing scale sizes $L = 2\pi/K = \lambda/\sin\theta$. Referring again to eqn. 5.15, r is a position vector, t is time and $f(r)$ is the delay function characterizing the surface. This complex function tells us how the scattering elements contributing to the 'bulk scattering cross-section σ' are distributed spatially.

Sea clutter background: characterization of ocean surface

By considering the inverse transform

$$f(r) \sim \int E(K) \exp(jK \cdot r) \, dK$$

we see that a general scattering medium can be characterized by a sum of Fourier components $E(K)$. For our electromagnetic field to give constructive back-scatter from this structure, the wave number of the electromagnetic wave K_{EM} must be 'matched to' (be equal to) that characterized by Fourier component of the reflecting structure K_{STR} (Crawford et al., 1959). We now dwell on this statement for a moment and consider the $f(r)$ function in relation to a reflector such as the sea surface.

First consider the case of a rigid (frozen) structure or a structure varying slowly with time. The general four-dimensional irregularity spectrum reduces to a three-dimensional spectrum

$$E(K) \sim \int_{-\infty}^{+\infty} f(r) \exp(-jK \cdot r) \, dr$$

Note that this simple treatment assumes a rough scattering surface (diffuse non-specular scattering). We first assume that the scattering surface is one-dimensional (unidirectional waves) in the sense that the lines of constant $f(r)$ are parallel. All waves, irrespective of wave length, are assumed to propagate in the same direction. This means that the delay function $f(r)$ reduces to a simple function of z normal to the wave front. The expression for the scattered field can then be written as (Gjessing and Irgens, 1964a and b)

$$E(K) \sim \int_{-\infty}^{\infty} f(r) \{\exp -j(K_x x + K_z z)\} \, dx \, dz$$

where x and z are orthogonal co-ordinates in the horizontal plane. However, with a set of unidirectional ocean waves, $f(r)$ reduces to $f(z)$, the scattering function (the delay function) being constant in the x-direction. Thus we have

$$E(K) \sim \int_{-\infty}^{\infty} \exp(-jK_x x) \, dx \int_{-\infty}^{\infty} f(z) \exp(-jK_z z) \, dz \quad (5.16)$$

The integral having x argument is immediately recognized as the Fourier integral representation of the Dirac delta distribution since $f(x)$ is constant, so that the above expression may be rewritten

$$E(K) \sim [\int_{-\infty}^{\infty} f(z) \exp(-jK_z z) \, dz] \, \delta(K_x x) \quad (5.17)$$

The scattered field $E(K)$ vanishes unless K_x is zero. Hence, $E(K) = 0$ unless K is normal to the wave crest. We see that in order for this statement to be strictly correct, we have to perform a $-\infty$ to $+\infty$ integration. This again implies that we have to illuminate the scattering surface over a large area compared to the ocean wave length under investigation.

Let us assume that we illuminate the sea surface with a beam whose width is limited so as to cut out a section D of the ocean wave front (see Fig. 5.6). In this case $E(K_x)$ will be of a sinc form instead of the Dirac delta function when approximating the antenna beam function to a rectangular distribution. Thus, if D is the width of the truncation influence $f(x)$, and $K = 2\pi/L$ is the wave number of the one-dimensional delay function $f(z)$ (L is the ocean wave length) then the angular width (angle between nulls) 2α of the $E(K_x)$ function is

$$2\alpha = \frac{L}{D}$$

This is a relationship well known from antenna theory. The beamwidth 2α (azimuth resolution) is determined by the antenna aperture D expressed in wave lengths L.

If the intensity distribution of the illuminator is not a simple rectangular one, the effect of the convolution (truncation) will be different. In general, we have to compute the Fourier transform of the illumination intensity in the x-direction in order to obtain the angular distribution of the radio power scattered back from a plane ocean wave. This scattering property of plane ocean waves can be illustrated in a simple manner. From Fig. 5.6 we see that if the radar antenna is pointing in a direction such that the phase front of the ocean waves coincides with that of the radio waves, constructive interference will take place. Turning the antenna through an azimuthal angle β, a situation arises where the radio waves which are scattered from one half of the illuminated ocean area cancel those scattered from the other half by destructive interference. This leads to a null in the angular power spectrum of the back-scattered waves. As seen from Fig. 5.6, the azimuth angle giving rise to zero back-scatter is given by $\beta = \tan^{-1}(L/2D)$. This is illustrated in Figs. 5.7 and 5.8 and verified experimentally in Fig. 5.26 (see Section 5.3). In Fig. 5.8 this function is plotted in polar co-ordinates.

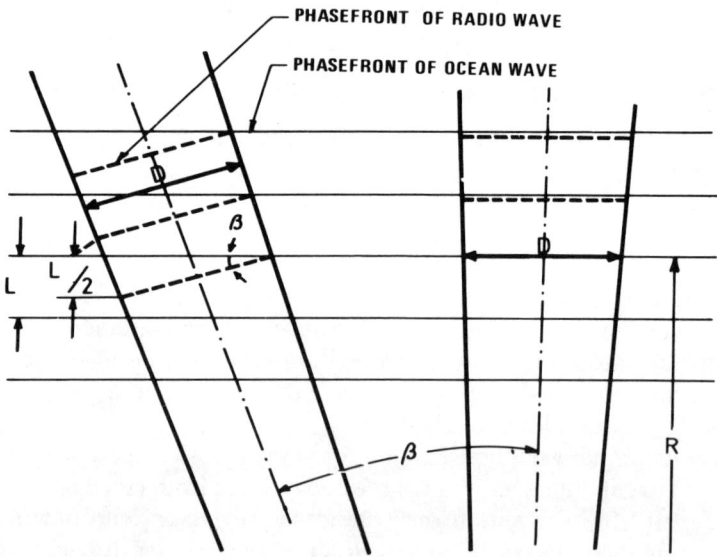

Fig. 5.6 Geometry of the truncation (convolution) process

Sea clutter background: characterization of ocean surface

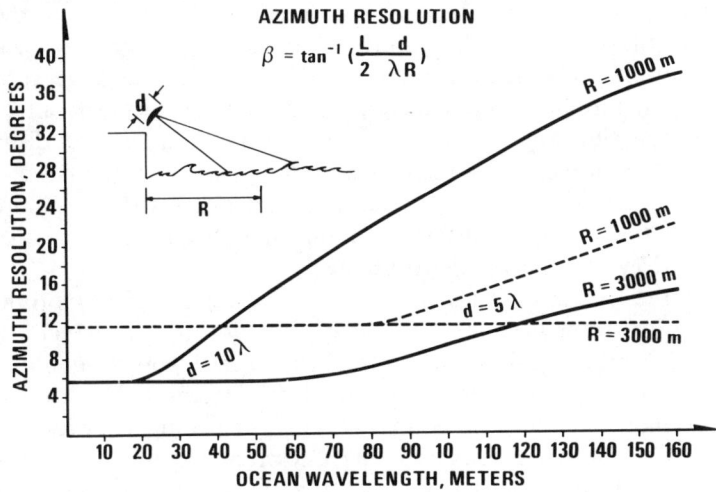

Fig. 5.7 Angular resolution is determined by ocean wave length and by the width of the illuminated spot on the sea surface

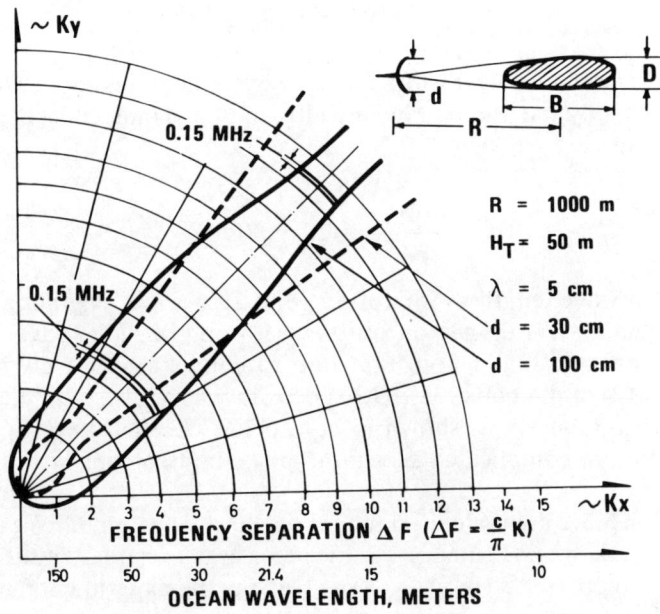

Fig. 5.8 Angular resolution of our radar varies with ocean wave length. As the ocean wave length decreases, the resolution approaches that of the antenna beamwidth
R is range; H_T is height of radar above sea surface; λ is radio wavelength, and d is antenna aperture (Gjessing and Hjelmstad, 1986)

Having considered the effect of a limited footprint size in regard to azimuthal resolution (wave direction), let us now consider the ocean wave length resolution. The radar antenna is illuminating a range interval B measured along the direction of wave propagation. This leads to a truncation of the delay function $f(r)$. We now assume that the field-strength distribution over the spot length B can be approximated by an exponential function. This is a reasonable assumption as the field strength decreases with the square of the distance and the intensity distribution through the antenna beam can be approximated by a Gaussian distribution.

The scattered field strength is, as we have seen (see eqn. 5.17), proportional to the Fourier transform of the delay function $f(r,t)$, which again is proportional to the incident field strength. Our exponential illumination distribution will, therefore, lead to a convolution integral. The sinusoidal sea surface is convolved with the exponential function leading to a widening of the filter function. If there had been no such truncation, our filter would have been a δ-function centred at the frequency $(c/2\pi L)$, where L is the ocean wave length. Note that we have disregarded the $\cos \alpha$ term stemming from a finite depression angle α. The exponential truncation leads to a widening of the filter function given by

$$\delta F = 0 \cdot 16 \frac{C}{B}$$

where δF is the half-power width (see, for example, Gjessing, 1978a and b). B is determined by the antenna beamwidth and by geometry (see Figs. 5.8 and 5.9). For this case we have

$$\frac{\delta F}{F_0} = 1 \cdot 28 \times 10^{-3} L$$

For a 100 m wave length we therefore get a 12% wave length resolution.

In concluding this discussion on the scattering of radio waves from plane ocean waves, we should note that due to convolution effects caused by a limited spot size the complex $E(K)$ spectrum will not be a simple expression involving a δ-function as shown in eqn. 5.17. We shall have to deal with a somewhat more complicated situation, an example of which is shown in Fig. 5.9.

In Section 5.3.2 a set of experimental results are presented. We now suggest a simple mathematical model based on basic and rather intuitive physical arguments to provide a simple way of assessing the experimental findings. We assume that long ocean waves (wave number approaching cutoff K_0) are unidirectional. The higher the wave number, the larger the angular spread. For waves in the wave number region above K_s, the irregularity structure is isotropic.

We shall not involve ourselves in a detailed discussion on angular spread as a

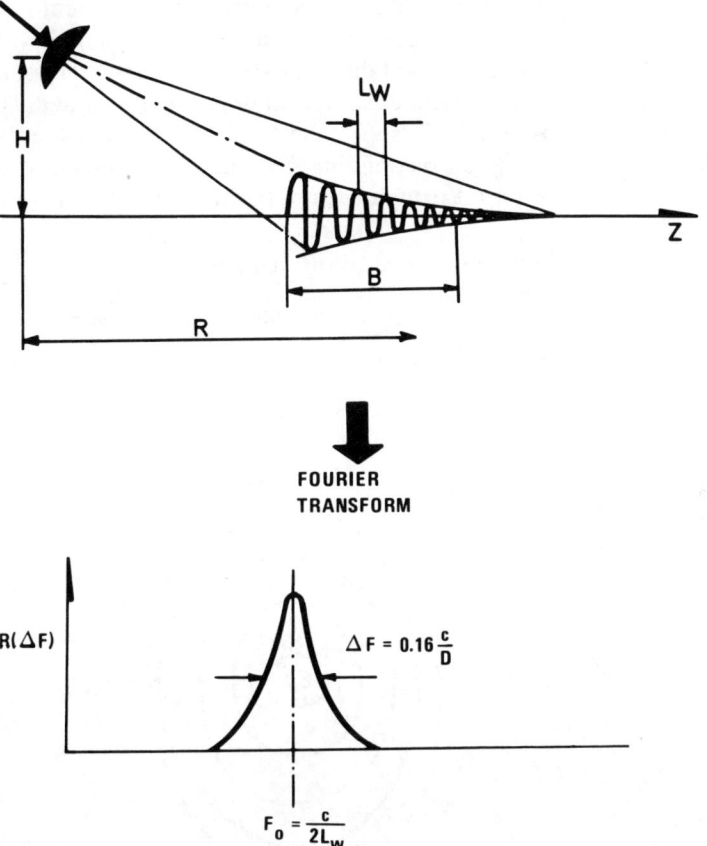

Fig. 5.9 As a result of a limited distribution in range of the illuminating spot, truncation effects (convolution) lead to degradation of the wave-number resolution (degradation of the coupling between electromagnetic waves and ocean waves)

function of wave number. We therefore write the complex irregularity spectrum simply as

$$E(\mathbf{K}) = E(|\mathbf{K}|, \theta) \sim K^{-n} (\cos \theta)^s$$

where s, the planeness parameter, is expressed as

$$s = \frac{K_s - K}{K - K_0}$$

K_0 is the wave number corresponding to the short wave number cutoff and K_s is the wave number at which the ocean surface structure becomes 'random' and isotropic with no prevailing wave direction. This form satisfies the limiting conditions expressed above: long waves are unidirectional since the exponent

$s \to \infty$ as $K \to K_0$. Short waves are isotropic since the exponent $s \to 0$ as $K \to K_s$. Fig. 5.10 shows an example of such a two-dimensional wave-number spectrum. Note that we have used the JONSWAP results of Hasselmann *et al.* (1973) for the scalar wave-number spectrum $\Phi_{JON}(|K|)$. For details the reader is referred to Phillips (1969). In short, the JONSWAP experiments show that the wave frequency f_0 corresponding to maximum spectral intensity is determined by the fetch F through the approximate relation $f_0 \approx F^{-1.4}$.

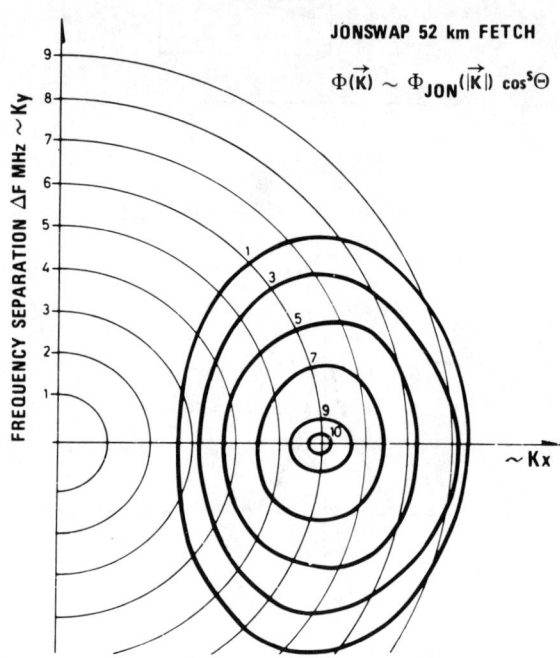

Fig. 5.10 *Complex wave spectrum calculated from the simple model presented above. Waves having long wave length are plane, small-scale irregularities are isotropic. The example presented is for $K_0 = 0$ and $K_1 = 2\pi/5$ m^{-1} (Gjessing, Hjelmstad, and Lund, 1985)*

Note that this $\cos^s \theta$ angular spread model gives a 50° half-power width of the angular spectrum of 25 m waves (6 MHz), while the 50 m waves have a spread of only 30°.

This simple model is in agreement also with the JANSWAP experimental results. By the aid of a linear array of wave meters, it was observed that the angular spread of waves in the vicinity of the cutoff frequency f_m is much smaller than that for shorter ocean waves.

As a means of comparison, the complex wave-number spectrum resulting from a $\cos^2 \theta$ directionality for all wave numbers is given in Fig. 5.11.

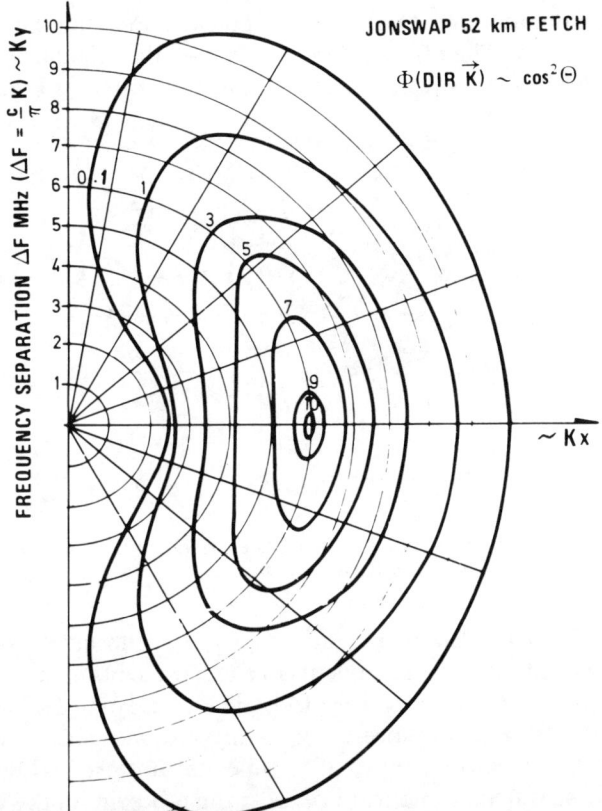

Fig. 5.11 *Complex wave spectrum based on a $\cos^2 \theta$ angular distribution and a JONSWAP one-dimensional spectrum*

The statement on angular spread in relation to ocean wave length can be motivated as follows:

1 Long wave length waves have a distant origin resulting from low attenuation. The larger the distance in relation to the 'source area' the more directional the wave. It takes a long fetch to produce a long wave length.
2 Long waves, once produced, react slowly to changes in local wind direction.
3 Short waves are produced at short ranges. Changes in the direction of local wind field have a strong influence on the short ocean waves.
4 Waves interacting with a local wind field give rise to a patchy gust pattern superimposed on the wave pattern. This 'fossilized turbulence footprint' (Bretherton, 1969; Woods, 1969; Lumley, 1967) is likely to give rise to incoherent motion as distinct from the well-behaved gravity waves. The symmetry axis of the wind gust pattern is likely to be centred on the direction of the mean wind. This does not necessarily coincide locally with the direction of the ocean waves.

A local wind field may, through shear mechanisms, break up the coherency of the ocean wave ridge. The effect of this is to widen the angular distribution of the back-scattered radar wave considerably, as illustrated in Fig. 5.12.

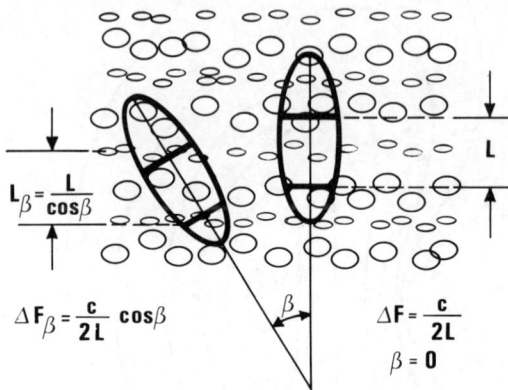

Fig. 5.12 *Scattering from turbulent irregularities caused by the 'footprint' of a local wind field interacting with the wave structure of the sea (Gjessing and Hjelmstad, 1986)*

It is well known from the turbulence theory that the nearness of a rough boundary, such as the sea surface, results in a low wave number suppression of the turbulence field, and, also, to anisotropy (see, for example, Panowsky and McCormick, 1960). Furthermore, any keen observer will have seen the modulation of the sea surface by wind turbulence. If there had been no wave structure on the sea surface one would expect the footprint of the wind field to give rise to an anisotropic irregularity structure on the sea surface with an axis of symmetry determined by the direction of the local mean wind (Gjessing, 1962). The presence of a wave structure on the sea surface will also modulate the wind field and seek to produce a sea-surface irregularity pattern with an axis of symmetry coinciding with that of the direction of the dominant wave spectrum. One would expect these factors to give rise to an incoherent irregularity structure on the sea surface with the 'axis of symmetry' along a direction somewhere between the direction of mean wind and that of the most energetic ocean waves (Gjessing, 1962; 1964). This is illustrated in Figs. 5.13 and 5.14.

Here we have assumed that the wave crests and troughs are broken up by turbulence and wave braking phenomena such that there is no 'phase matching' between radio wave and ocean wave over the illuminated area, contrary to the coherent case depicted in Fig. 5.6.

Thus, when turning the antennas an azimuth angle β, the radio wave sees a dominating ocean irregularity scale which is increased from L when $\beta = 0$ to

$$L_\beta = \frac{L}{\cos \beta}$$

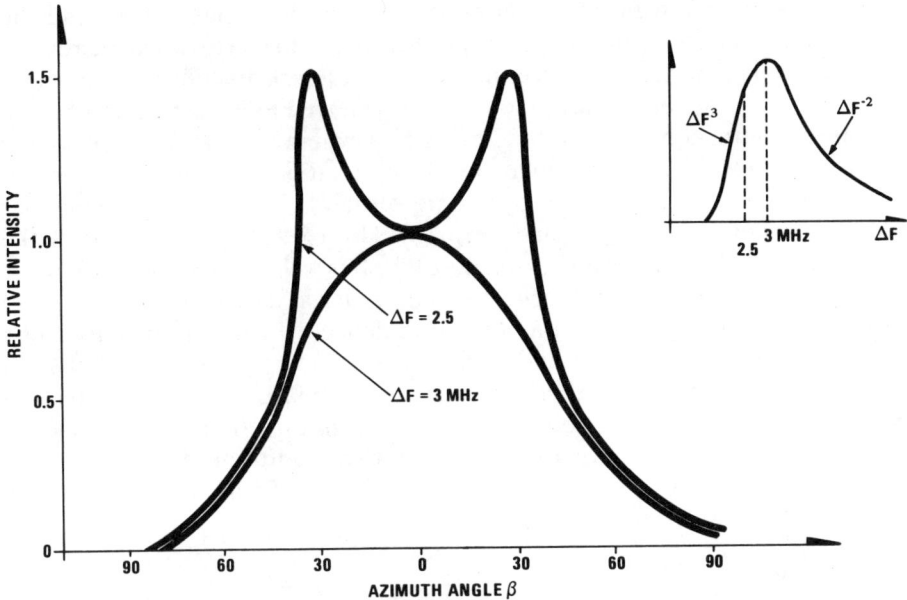

Fig. 5.13 *Scattering from turbulent irregularities is expected to give a two-lobed angular response for frequencies below the spectral peak (Gjessing an Hjelmstad, 1986)*

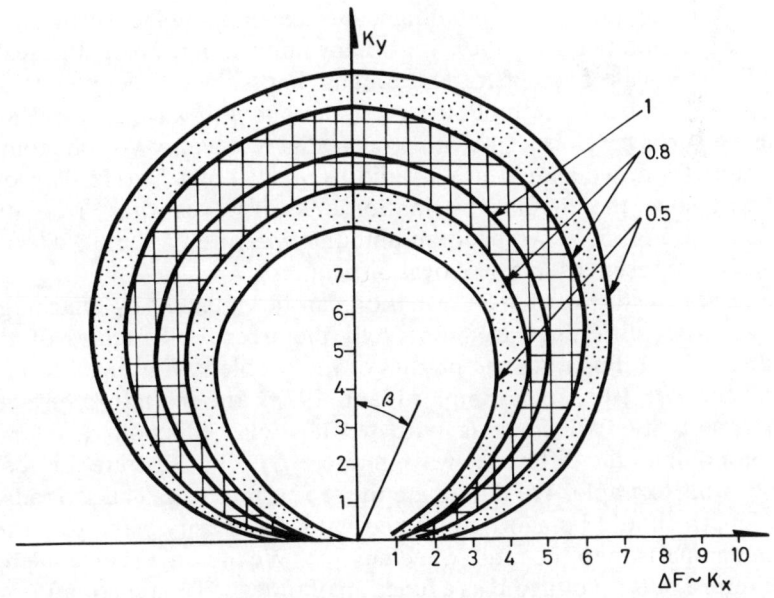

Fig. 5.14 *Directional wave number plot of scattering from turbulent irregularities (Gjessing and Hjelmstad, 1986)*

This is illustrated in Fig. 5.12. Note that for certain values of K (and the corresponding ΔF) the back-scattered radio signal will increase with increasing azimuth angle symmetrically on either side of the $\beta = 0$ direction, see Fig. 5.13. For experimental verification, the reader is referred to Figs. 5.30 and 5.32.

This phenomenon is illustrated in Fig. 5.14 where the angular relationship above is plotted assuming the following $\Phi(|k|)$ relationships: the wave-number spectrum can be approximated by two power laws. On the low wave number side of maximum intensity it can be expressed by a K^m relationship, and on the high wave number side as a K^{-n} relationship. It is well known from turbulence theory that the nearness of a rough boundary, such as the sea surface, results in a low wave number suppression of the turbulence field and also in anisotropy (see, for example, Panofsky and McCormick, 1960). (Modulation of the sea surface by the wind turbulence, anisotropic irregularity structure on the sea surface with an axis of symmetry determined by the direction of the local mean wind (Gjessing, 1962).) In the case of a wide footprint (large coherence distance along the wave crest) the directional pattern will, obviously, be the same as in the coherent case.

5.3.1 Multifrequency radar principle and frequency difference matching

Up to this point we have been dealing with electromagnetic waves having wavelengths which are comparable with the irregularity scale of the scattering surface. We saw that by illuminating a surface with an electromagnetic wave having wave number $K = 4\pi/\lambda$ we obtain information about the scattering structure at a scale $L = 2\pi/K$. This means that we are directly matching the radio wave number K to the Fourier component with wavenumber $2K$ of the scattering surface. Thus, if we are to investigate ocean wave phenomena of scale, say 100 m, we should use a radio frequency of 1·5 MHz. However, in order to resolve the angular distribution of such waves with a resolution of, say, 10° (see Fig. 5.7), we need an antenna aperture of some 1200 m. This, obviously, represents some practical difficulties.

We now investigate what information can be obtained by matching some beat patterns (difference frequency) to the irregularity scales of the sea surface. Fig. 5.1 illustrates the physics of the problem. For this ΔK matching scheme to work (see, for example, Plant, 1977), the scattering surface must have irregularities which contribute to the delay function $f(z)$ for scales corresponding to the difference wave number ΔK. To illustrate this, consider the following example. We illuminate the sea surface with a pulsed radar. The pulse length should be short compared with the gravity wave structure and long in comparison to the radio wavelength λ. We measure the detailed shape of the return pulse (voltage V as a function of time τ). The function $V(c\tau)$ gives us, directly, the delay function $f(z)$. If the gravity wave structure shows up on the delay function $f(z)$, i.e. if the spectrum of $f(z)$ has Fourier components matching the wave number ΔK, we would experience a radar return at the beat frequency $\Delta\omega = c\Delta K$.

There are many ostensible explanations for this 'modulation' of the small-scale structure (to which the wave number K is matched) by the long wave phenomena (to which ΔK is matched) see, for example: Plant, 1977; Plant and Schuler, 1980; Clifford and Barrick, 1978; Jones and Weissman, 1981; Weissman and Johnson, 1977; Weissman et al., 1982; Valenzuela, 1978a and b; Wright et al., 1980).

1. It is well known that the scattering cross-section of a surface increases with incidence angle. This means that from a sinusoidal gravity wave we get maximum low angle return at the point of maximum slope, i.e. from the inflection points where the second derivative is zero.
2. The turbulent intensity is governed by damping factors such as shear forces (Woods, 1969; Lumley, 1967). Orbital motion within the wave structure changes the velocity shear periodically in phase with the gravity wave.
3. Shadowing effects may enhance the patchiness of the radar illumination and strengthen the apparent modulation effect of the gravity waves (Bass and Fuks, 1979).
4. When the radius of curvature of the wave surface is the same as the distance from the radar to the scattering region in question, we may experience focusing (Gjessing, 1964; Gjessing and Irgens, 1964a and b).

Breaking waves and white capping resulting from the local wind field etc., complicates the issue still further. (See Fig. 5.15.)

In this presentation on the physics of electromagnetic interaction mechanisms, we make no attempt to resolve the question of scattering cross-section (delay function $f(z)$) in relation to wave height in a quantitative manner. We constrain ourselves to giving qualitative results and note that empirical relations and also promising theoretical models are about to become available (see, for example, Valenzuela, 1978a and b; Plant, 1977; Schuler, 1978).

Note that there is nothing new about the characterization of scattering surfaces by *bandwidth measurements*. As early as in 1961 such measurements were performed by Alan Waterman at Stanford University (Waterman, Gjessing and Liston, 1961). Also pioneers in this field were Crawford, Hogg and Kummer from Bell's Research Laboratory (1959). The Stanford group, as well as that from Bell's, focused their attention on wave phenomena in the troposphere superimposed on which there are also incoherent turbulence, much in the same way as for the ocean surface with coherent gravity waves and capillary waves with turbulence. This problem was also addressed by Tor Hagfors as early as in 1959. His approach is indeed very analogous with the current one. Instead of investigating the scattering properties of a sea surface consisting of small scale 'incoherent' irregularities riding on large scale wave phenomena, Hagfors studied the layered/turbulent structure of the ionosphere from below making use of the bandwidth properties of the scattering medium (Hagfors, 1959). A very similar technique was used by Hagfors at Stanford

University to measure the topography of the lunar surface (Hagfors, 1961). Here the 'delay spectrum' was deduced from measurement of the correlation function in the frequency domain, see also Ishimaru (1978) and Tatarsky (1961). Conceptually, therefore, the problem of ΔK matching to the sea surface is very straightforward and does not require a sophisticated consideration of capillary resonance phenomena. Any scattering surface, such as a ploughed field, having iregularities of scale K coupling to the carrier electromagnetic wave and also scales ΔK coupling to the beat frequency wave, lends itself to multifrequency investigations (ΔK matching).

For this ΔK matching scheme to work, the scattering surface obviously must have irregularities which contribute to the delay function $f(z)$ for scales corresponding to the difference wave number ΔK. If the gravity wave structure shows up in the delay function $f(z)$, i.e. if the spectrum of $f(z)$ has Fourier components matching the wave number ΔK, we would experience a radar return at the beat frequency $\Delta\omega = c\Delta K$.

As an illustration emphasizing the physics of the problem, consider the following question which is currently being addressed by the authors' organization. One is looking for a method by which a forest of Norway spruce can be characterized by the multifrequency matched illumination method. The planted spruce forest consists of a grid of trees. There are many dominating scales, like on the sea surface, to which K and ΔK shall have to be coupled.

Obviously a carrier frequency should be chosen so as to give resonant coupling to the needle dipoles. The length distribution of the ensemble of needles is very narrow and centred round approximately 20 mm. For these to act as efficient dipoles, the carrier frequency should be approximately 7·5 GHz (to couple to a Scotch pine, the carrier frequency should be approximately 3 GHz). Then the branch of the spruce tree consists of sub-shoots which typically are 8 cm long, whereas the length of the branch itself could be 1 m and the vertical spacing of branches something like 1/2 m.

If the interest now is limited to the measurement of the distances between the rows of treas (typically 2–5 m) we should use a carrier frequency of 7·5 GHz and difference frequencies in the range from 30 to 75 MHz. The antennas should as in the case of the sea investigation, be directed nearly horizontally and in an azimuth direction which corresponds to the direction where the distance distribution is sought.

In this context, obviously, it is not meaningful to use the term 'modulation transfer function'; neither is it perhaps appropriate to refer to 'Bragg scattering'. The mathematics of difference frequency matching are given in Section 2.1.

Before bringing this section on matched illumination in relation to ocean waves to a close, it may serve a useful purpose to present a very brief overview of other powerful techniques for sea surface radar investigations. The more important of these techniques are illustrated in Figs. 5.16–5.21 (see also Jackson, 1980; Alpers and Rufenach, 1979).

Sea clutter background: characterization of ocean surface

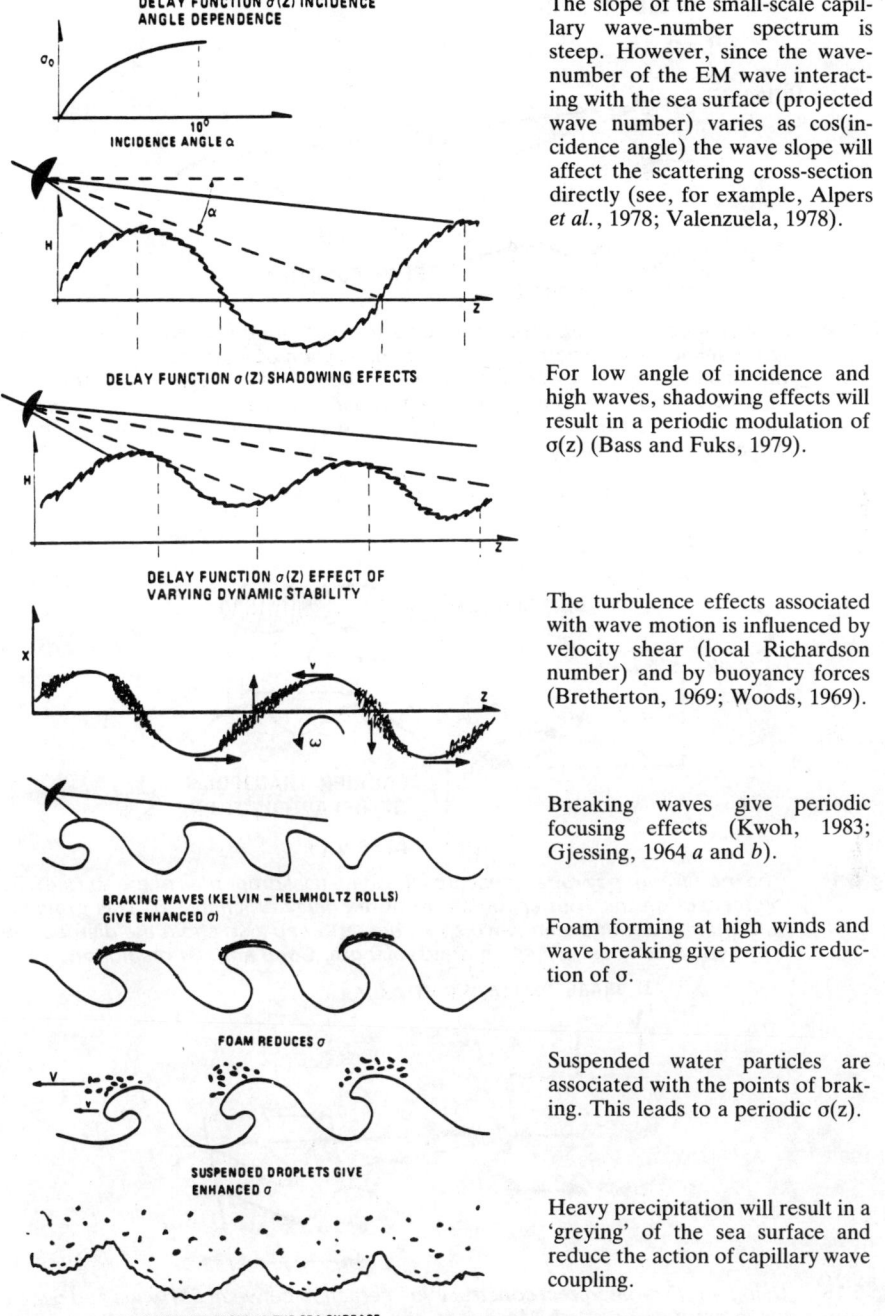

The slope of the small-scale capillary wave-number spectrum is steep. However, since the wave-number of the EM wave interacting with the sea surface (projected wave number) varies as cos(incidence angle) the wave slope will affect the scattering cross-section directly (see, for example, Alpers et al., 1978; Valenzuela, 1978).

For low angle of incidence and high waves, shadowing effects will result in a periodic modulation of $\sigma(z)$ (Bass and Fuks, 1979).

The turbulence effects associated with wave motion is influenced by velocity shear (local Richardson number) and by buoyancy forces (Bretherton, 1969; Woods, 1969).

Breaking waves give periodic focusing effects (Kwoh, 1983; Gjessing, 1964 a and b).

Foam forming at high winds and wave breaking give periodic reduction of σ.

Suspended water particles are associated with the points of braking. This leads to a periodic $\sigma(z)$.

Heavy precipitation will result in a 'greying' of the sea surface and reduce the action of capillary wave coupling.

Fig. 5.15 *Various mechanisms giving a periodic distribution of scattering cross-section with range, which is controlled by the dynamics of the sea*

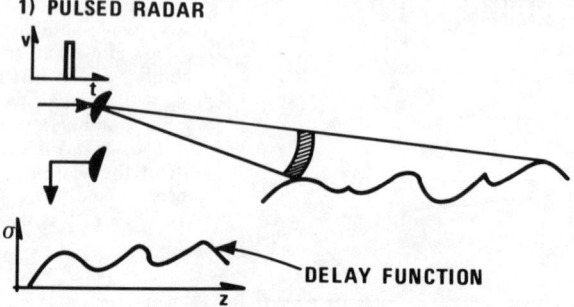

Fig. 5.16 *A conventional pulsed radar measures the distribution in range of scattering cross-section as a 'snap shot' (i.e. one realization of a statistical ensemble) if the pulse width is small compared with the ocean wave length. Poor azimuthal (angular) resolution. Using a coherent radar with a stationary resolution cell small compared to the wave length, improved directional resolution can be achieved (see Fig. 5.21)*

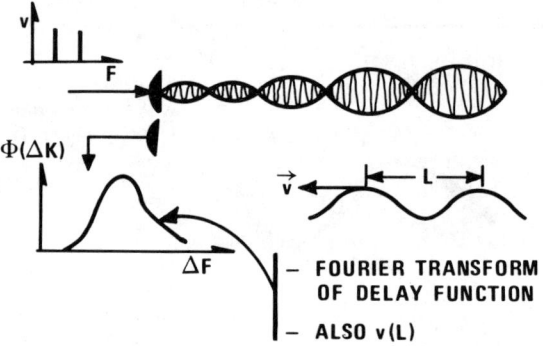

Fig. 5.17 *The multifrequency radar measures the statistical properties of the sea surface by measuring the Fourier transform of the delay function $\sigma(z)$. This provides information about ocean wave spectrum $\Phi(K)$ and also about the distribution of wave velocity as a function of wavelength. Good angular resolution*

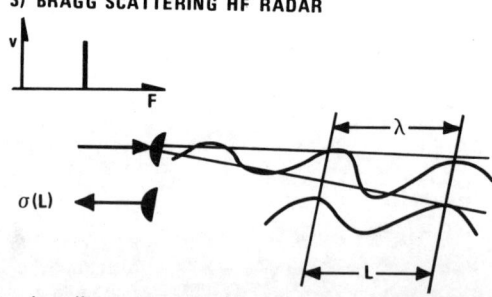

Fig. 5.18 *Using an HF radar, direct constructive interaction between EM wave and ocean wave can be achieved. Measuring the Dopler shift information about wave direction can be obtained on the basis of knowledge about wave dispersion relationship. Varying the EM frequency over the appropriate frequency band, $\sigma(L)$ can be obtained. Poor angular resolution*

4) SAR

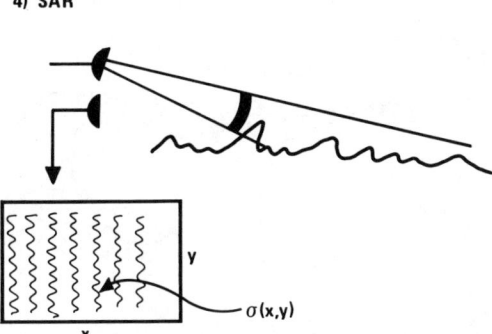

Fig. 5.19 Synthetic aperture radar SAR is based on a short pulse (FM chirp, pulse compression technique, etc.) for range resolution. For azimuthal resolution the principle of synthetic aperture is made use of. This involves the measurement of amplitude and phase as the flying radar sweeps its beam over the scattering surface

5) RADAR ALTIMETER

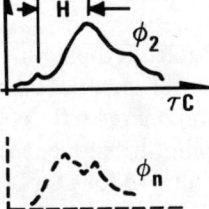

Fig. 5.20 Radar altimeter makes use of a short pulse. Measuring the rise and fall time of the back-scattered pulse pattern as the pointing direction of the antenna beam is changed, direct information about wave height and wave direction is achieved. Note that this experiment does not require information about the relationship between wave height and scattering cross-section (Welsh et al., 1979)

116 *Sea clutter background: characterization of ocean surface*

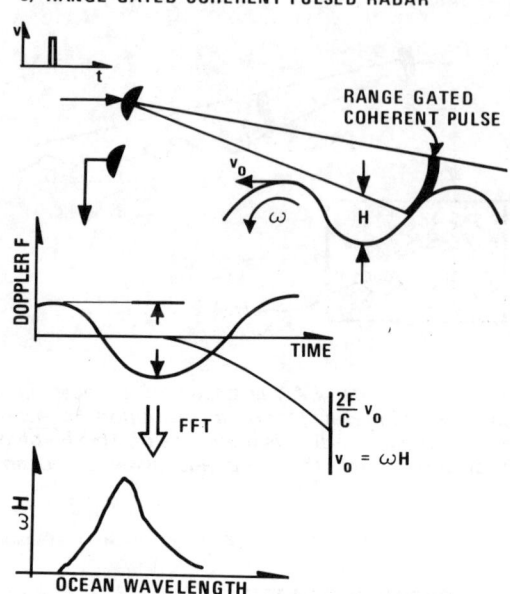

Fig. 5.21 *Range-gated coherent pulsed radar involves a pulse to pulse coherent transmitter. Measuring the frequency modulation as a function of time of the back-scattered signal, the orbital velocity is obtained through Doppler considerations when the pulse width is small compared with the ocean wave length. Since the phase velocity v = ωH information about wave height is obtained under ideal sea conditions (Grønlie, Brodtkorb, and Wøien, 1984)*

In summing up this section on the characterization of sea surface by modern radar methods, Table 5.1 may serve a purpose.

5.3.2 Some illustrative experimental examples

Having laid the foundation for an understanding of the principles involved in relation to the coupling of electromagnetic waves to ocean waves (ΔK coupling), some illustrative examples may strengthen our physical understanding. Before we present the experimental arrangements and give some illustrative results, however, it is of importance to perform some simple calculations so as to have some idea of what to expect from the experiments.

Basing our calculations now on the expression for Doppler shift given earlier (see Section 5.1)

$$f = \sqrt{\frac{2\Delta F}{\pi c}} \qquad (5.17)$$

and on the familiar JONSWAP spectrum for ocean waves referred to above, we can calculate the Doppler shift and spectral intensity associated with each radio frequency pair subject to the assumption that we know the relationship

Table 5.1 *Characterization of ocean surface (gravity waves, internal waves, wind pattern, ship wakes). The relative potential of modern radar systems*

	Adaptive multi-frequency	Pulse compression	Chirp FM/CW	SAR
1-D spectrum $A(\|K\|)$	yes	yes	yes	yes
2-D spectrum $A(\|K\|, \omega)$ space/time	yes	limited	no	no
2-D spectrum $A(\|K\|)$ 2-D space	yes	yes	limited	yes
3-D spectrum $A(K, \omega)$ velocity spectra as a function of scale size (coherent/non-coherent)	yes	no	no	no

Matched illumination for specific wakes
  Directional spectra for non-dispersive phenomena
  Directional spectra for dispersive waves
	yes	no	no	no

Space/time coherence for different scales	yes	no	no	no

between scattering cross-section and wave height. Fig. 5.22 relates to the multifrequency radar concept. We select a given ΔF_1 (which is the same as selecting a given irregularity scale L_1), and we measure how the frequency covariance function

$$R(\Delta F_1, t) = V(F,t) V * \{(F + \Delta F_1), t\}$$

is varying with time t. We then compute the power spectrum of the frequency covariance function $R(\Delta F_1,t)$. This means, as depicted in Fig. 5.22, that we apply a Fourier transformation of the $R(\Delta F_1,t)$ function with respect to time t so as to obtain the Doppler spectrum. If, then, the ordered gravity waves dominate over the incoherent velocity components which ride on the gravity waves, we would expect the Doppler spectrum to have a maximum at the frequency corresponding to the dispersion relation of gravity waves (eqn. 5.17), and we would expect a Doppler broadening determined by the velocity spread δV. Hence, the Doppler broadening Δf is given by

118 Sea clutter background: characterization of ocean surface

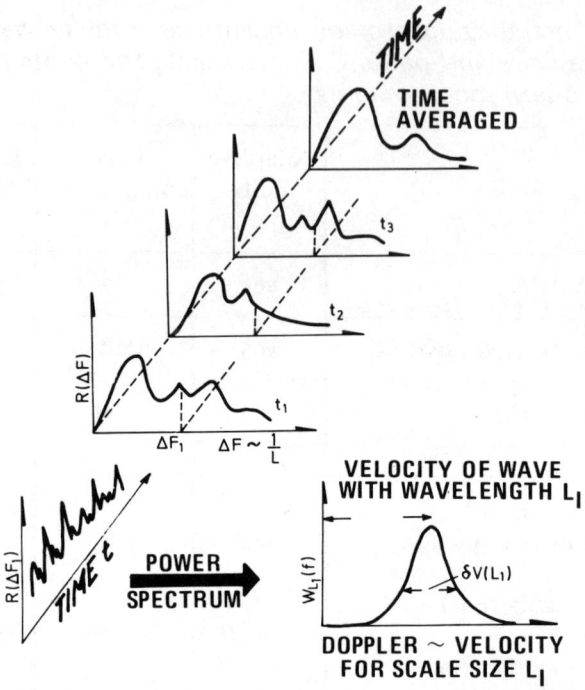

Fig. 5.22 *By studying the temporal variation of the frequency covariance function $R(\Delta F)$ for a given frequency ΔF_1 scale size $L_1 = c2\Delta F_1$) we derive information about the motion pattern (Doppler) of the irregularity scale L_1 (Gjessing, 1978a and b)*

$$\Delta f = \frac{2\Delta F}{c}\delta V \qquad (5.18)$$

The detailed signal analysis procedure is shown in Fig. 5.23.

The following calculations will be based on the JONSWAP experimental results shown in Fig. 5.5. Knowing from this figure the ocean wave-height spectra in the conventional time domain, it remains, before we can calculate the radar signature, to convert the temporal spectra to spatial spectra, applying the relationships given in Section 5.1 (frequency $f = \sqrt{(g/2\pi L)}$ where L is the ocean wave length). Thus the frequency axis of Fig. 5.5 showing the JONSWAP spectra is converted directly to a beat frequency ΔF axis by the relationship

$$f = \sqrt{\frac{g\Delta F}{\pi c}}$$

This is shown in Fig. 5.24

Then, finally, let us present some illustrative experimental results.

(a) Directional ocean wave spectra observed from the NASA Electra aircraft
Illuminating the sea surface by a fixed 14° beamwidth side-looking antenna pointing 10° downwards relative to the horizontal plane, the ocean wave

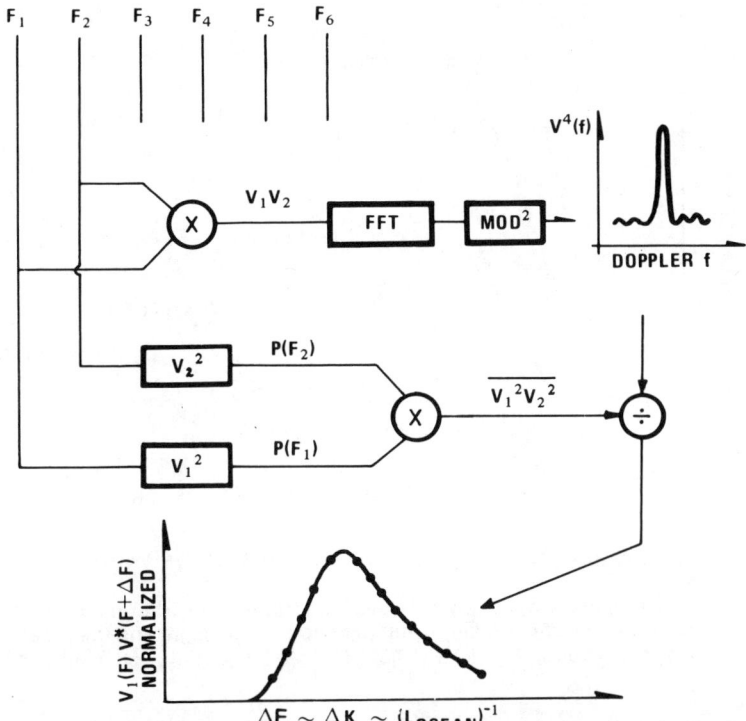

Fig. 5.23 *Multiplying the various combinations of the six frequencies, 15 values of ΔK are obtained, coupling to 15 ocean scales. Thus we get information about the motion pattern of 15 ocean scales (coherent and incoherent components) and we also get information about the scattering cross-section through the Fourier transform of the delay function $\sigma(z)$. The spectra are normalized as indicated in the figure*

spectra were determined for various azimuth directions and for 15 different ocean wave lengths in the interval from some 16 m (couples to $F = 8$ MHz) to 300 m (corresponding to $\Delta F = 1/2$ MHz). For each frequency separation ΔF the frequency covariance function $V(F,t)V*(F + \Delta F,t)$ was computed.

The NASA Eelctra operated north-east of Wallops Flight Facility (38.49.2 N–74.24 W) at an altitude of 12 000 ft, 20th January, 1983, time 2203Z.

Banking the aircraft 10° so that the antennas were pointing at a depression angle of 20°, the aircraft completed a 360° circle. In this way the antenna was illuminating essentially the same area on the sea surface at all azimuth directions. The longitudinal dimension of the footprint is approximately 5600 m, whereas the transverse dimension is approximately 2000 m.

Flying in a closed circle, the frequency covariance function $V(F)V*(F + \Delta F)$ was computed for 15 different values of ΔF for all azimuth angles. The results are shown in Fig. 5.25. Note that the Bragg angle is clearly visible for the longer wave lengths (215–88 m).

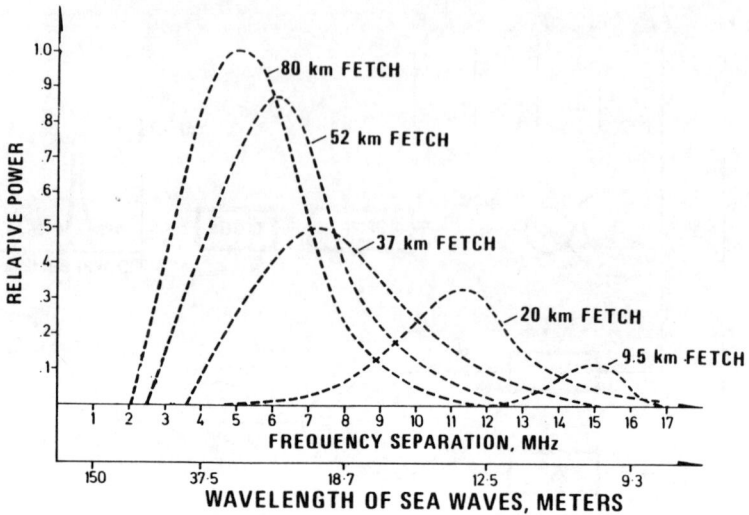

Fig. 5.24 ΔK signature of the ocean surface based on the JONSWAP experiments of Fig. 5.5
Note that the relative power (wave height) axis is in arbitrary units. There is not necessarily a linear relationship between wave height and the measured radio quantity given in Fig. 5.23 $V(F)V * (F + \Delta F)$ (Gjessing, 1981a and b © IEEE)

In terms of aircraft altitude H, depression angle α and antenna beamwidth $\Delta\beta$ we get the following expression for the Bragg angle

$$\beta = \frac{L}{2H} \frac{\sin\alpha}{\sin(\Delta\beta/2)}$$

This is presented in Fig. 5.26. Note the good agreement between theory and experimental results.

By noting the two azimuth angles corresponding to destructive interference on either side of the wave direction, this can be determined with great precision, as shown in Fig. 5.27.

Finally, for an azimuth direction corresponding to that of the propagation direction of the ocean waves, the distribution with wave length of the scattering cross-section is computed. This is shown in Fig. 5.28. Note that two wave lengths dominate, 130 m and 18 m.

(b) Directional spectra obtained from ground-based station on a cliff
The complex field strength was recorded over a period of some 12 minutes before the azimuth angle was changed, thus obtaining a plot of power spectral density and Doppler spectrum as a function of frequency separation (ocean wave length). A typical result is shown in Fig. 5.29.

Sea clutter background: characterization of ocean surface

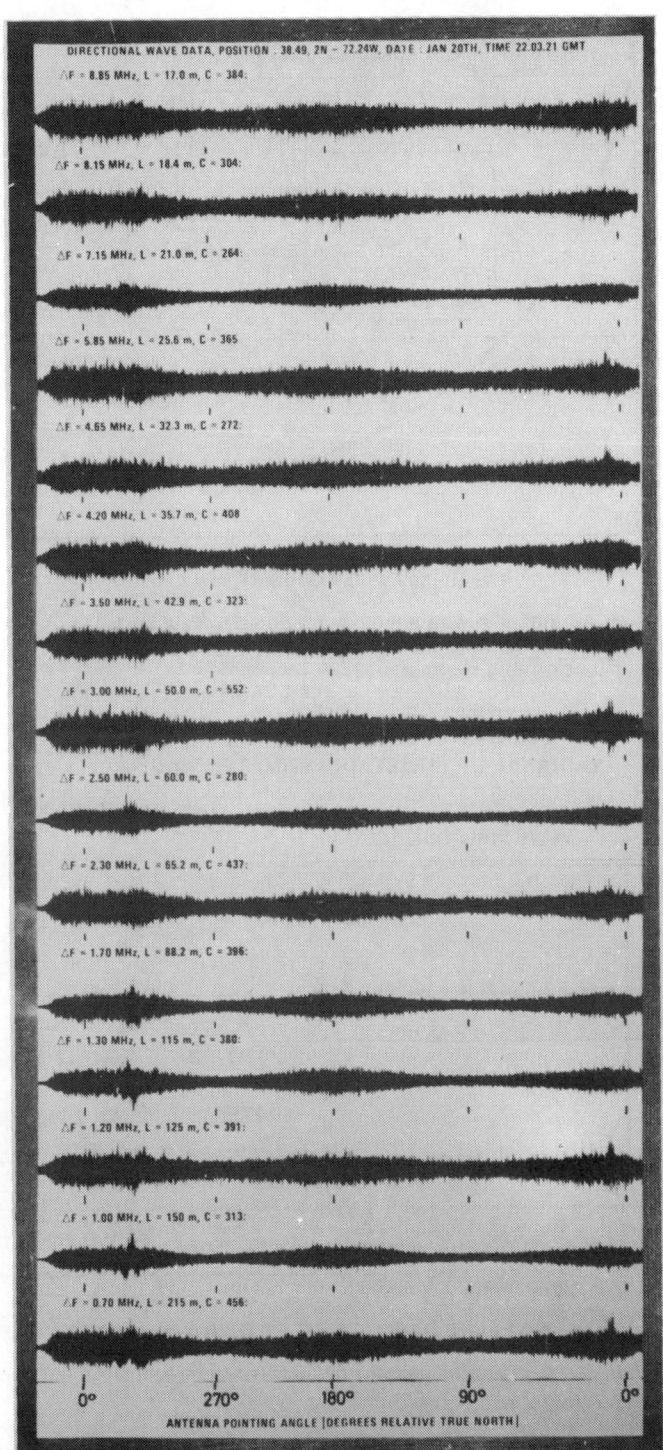

Fig. 5.25 *The aircraft flew in a closed circle pointing the antennas 30° down from 12 000 ft. The covariance function $V(F)V * (F + \Delta F)$ is shown for 15 different values of ΔF (Gjessing and Hjelmstad, 1986)*

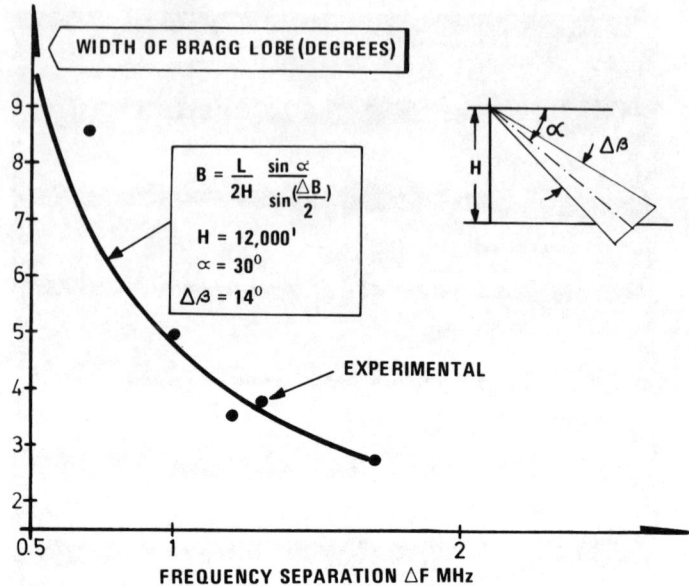

Fig. 5.26 *Measurement of Bragg angle is compared with theory*
Note that the aircraft altitude is 2000 ft, depression angle 30° and beamwidth 14° (Gjessing and Hjelmstad, 1986)

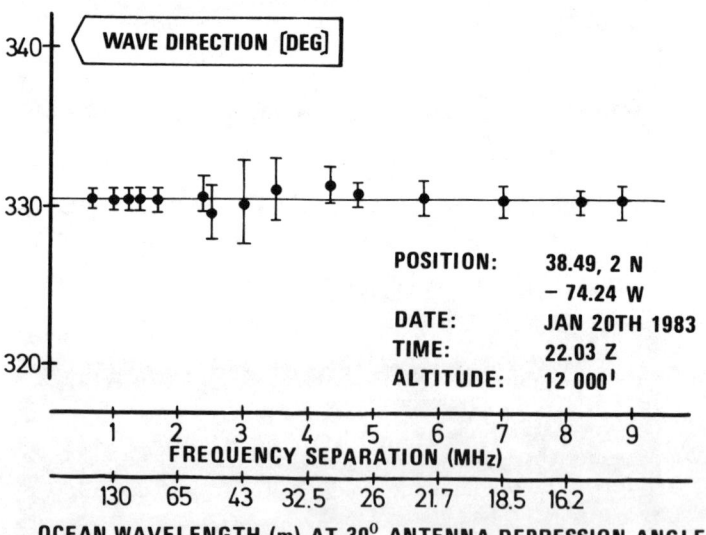

Fig. 5.27 *Wave direction plotted as a function of wave length (Gjessing and Hjelmstad, 1986)*

Sea clutter background: characterization of ocean surface

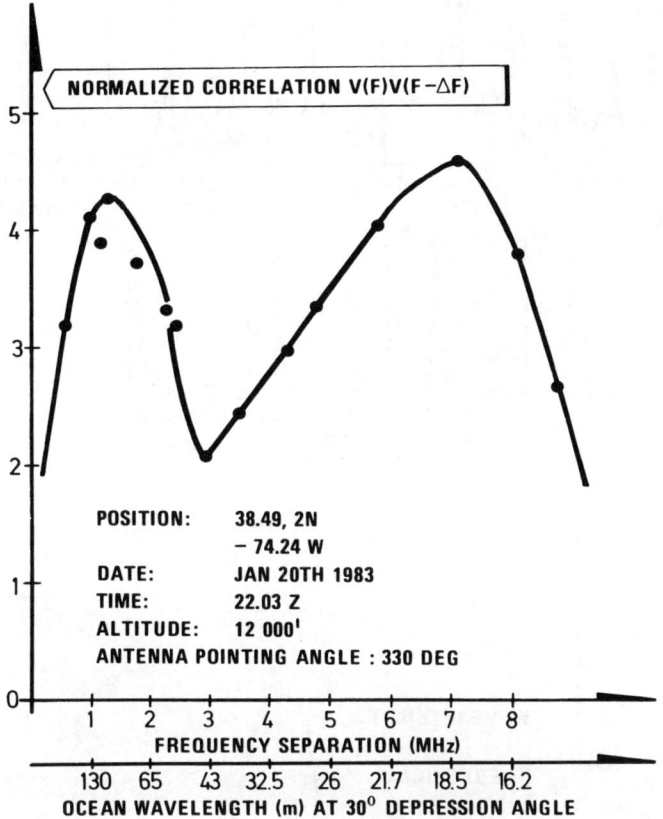

Fig. 5.28 *There are two dominating wave lengths, 130 m and 18 m*

First, consider the results obtained on 12th March, 1981 (1010–1235 local time). The weather and sea state can be characterized as follows:

Wind direction 90° (east)
Wind speed approximately 8 m/s
Wave height 1·4 m peak-to-peak
Waverider gives maximum wave height for $f = 0·225$ Hz
Wind decreasing at noon, turning northerly. Wave height reduces to 1 m peak-to-peak, choppy sea

Fig. 5.29 shows the wave intensity for a particular azimuth direction and also the Doppler spectra associated with various scale sizes. Note the very well defined spectral peak in the intensity versus wave length distribution. Note also that the Doppler frequency is in very good agreement with the theoretical dispersion relation. There seems to be a slight, but consistent, displacement to higher Doppler frequency. This is a result of an inward current.

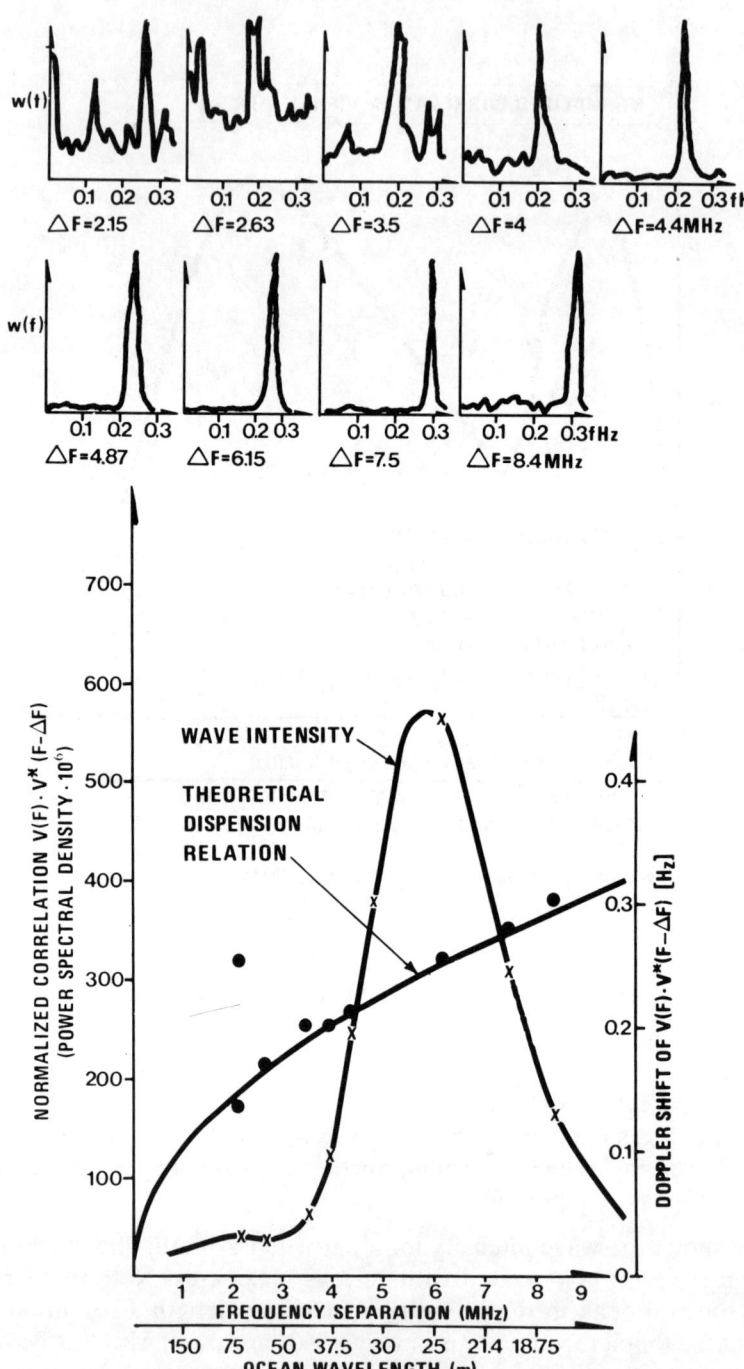

Fig. 5.29 *Spectra of wave length and wave velocity*
Note that wave lengths shorter than that corresponding to the spectral peak move at constant velocity. This is not the case for longer scales. The results are from Langesund, 12th March, 1981, at 1030 local time, the azimuth direction being 135° (Gjessing *et al.*, 1985)

Sea clutter background: characterization of ocean surface 125

Finally, in Fig. 5.30, we present the two-dimensional irregularity spectrum summing up the description of the sea-surface structure. Here we have plotted the isolines of spectral intensity, or more specifically the quantity $V(F)V * (F + \Delta F)/V(F)^2$, in the K_x–K_y plane. This presentation should be compared with those of Figs. 5.10, 5.11 and 5.14. Note also that the isolines of spectral intensity $\Phi(\Delta K)$ are obtained directly from a set of one-dimensional spectra such as those shown in Fig. 5.29.

As will be seen, we distinguish between coherent and incoherent components. With 'coherent component' we refer to the dominant Doppler spectral line caused by the 'coherent' gravity wave ($f = \sqrt{\{g\Delta F/\pi c\}}$). The incoherent component is the 'residue intensity' of the Doppler spectrum (the contribution from the skirts) obtained when the coherent contribution wave is subtracted. It can be visualized as all the random turbulent velocity contributions outside the short velocity interval dominated by the gravity wave. Note, however, that although the velocities of the scatterers vary widely, as seen, for example, from

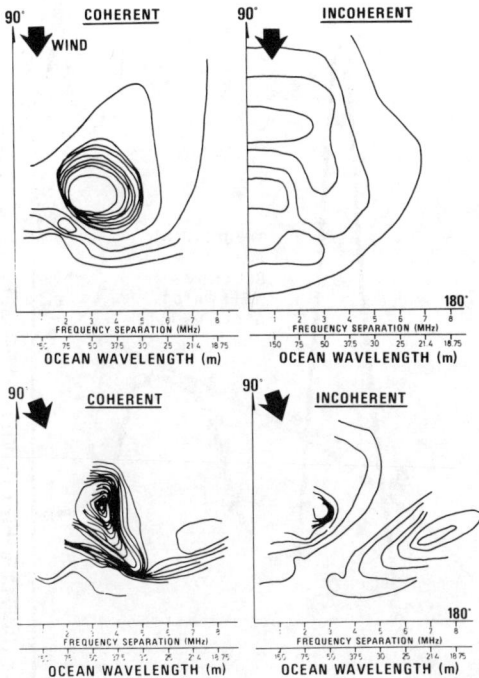

Fig. 5.30 *Isolines of spectral intensity $\Phi(\Delta K)$ in the K_x–K_y plane*
Note the effect of changes in wind direction on the coherent irregularity spectrum. Note that the upper set of maps was obtained on 12th March, 1981, 1010 to 1135 local time, whereas the lower set refers to the situation during the time interval 1135 to 1235 (see also Gjessing, Hjelmstad, and Lund, 1985)

126 Sea clutter background: characterization of ocean surface

Fig. 5.31, the difference frequency ΔF selects a narrower range of irregularity scales centred around the scale size L given by

$$L = \frac{c}{2\Delta F}$$

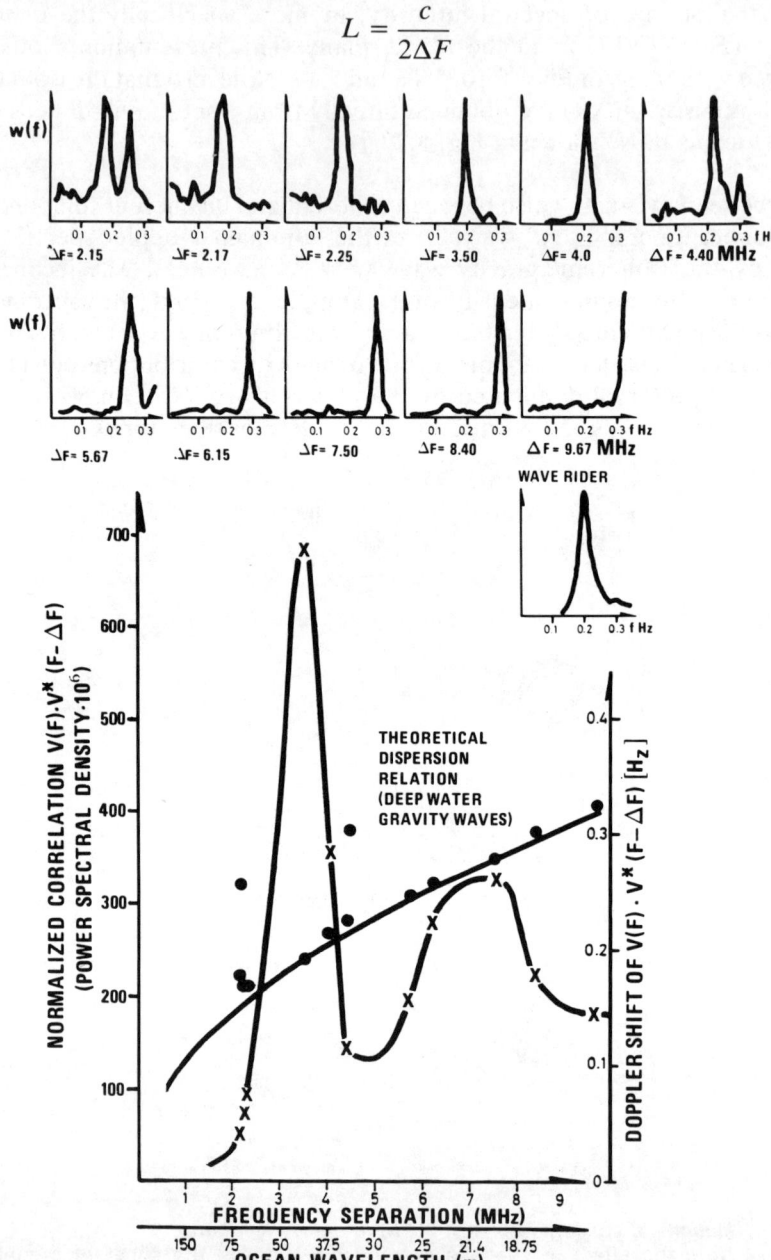

Fig. 5.31 *Spatial and temporal wave spectra for an azimuth direction of 140°
Location is Langesundsfjorden, time is 18th March, 1981, 1525 local time*

If these weak incoherent irregularities were entirely due to the local wind field, one would expect the local wind direction to provide an axis of symmetry. Unfortunately, the antenna systems were not flexible enough to be turned beyond the east direction ($\theta = 90°$). It is therefore difficult to establish such a symmetry. The coherent contribution, however, has a pronounced axis of symmetry determined by the direction of the most energetic ocean wave. We see that the angular distribution is far from $\cos^2 \theta$ distributed, long waves are more plane than the shorter ones, as suggested in Fig. 5.11.

The lower set of directional spectra in Fig. 5.30 gives the situation one hour later when the wind has turned from east towards north. We see that the most energetic waves (the 'eye' of the isoplot chart) have not changed direction, whereas the shorter scales and weaker disturbances are reoriented slightly. The distribution of the incoherent (turbulent) irregularities is drastically altered.

On 18th March, 1981 there was no wind before noon. At 1330 the wind speed was 10 m/s from the south, bringing the wave height up to 2·5 m. The complex K plot shows a unidirectional spectrum.

Two hours later, when the wave height increased to 2·5 m peak-to-peak and the wind turned easterly, the irregularity structure was very much altered. This most striking feature had (as indicated in Fig. 5.31 and also summarized in Fig. 5.32) the very pronounced appearance of a second peak at a wave length of 25 m, whereas the dominating spectral peak was still 50 m. Note that this two-peak spatial spectrum for a particular azimuth direction is not very pronounced in the wave rider spectrum. This, presumably, was a result of the omnidirectional response pattern of the wave rider. Note also that Fig. 5.31 shows a very pronounced two-peak structure in the Doppler spectrum for an ocean wave length of some 70 m. Finally, Fig. 5.32 sums up the results of the afternoon run in the form of $\Phi(\Delta K)$ isolines for the coherent as well as for the incoherent components. Note that the wave length corresponding to peak spectral intensity varies with azimuth angle.

For details regarding sea-surface signatures in relation to the matched illuminator concept, the reader is referred to Gjessing, Hjelmstad, and Lund (1985) and Gjessing and, Hjelmstad (1986).

Before ending this section on illustrative examples in relation to the sea surface, it may be of interest to compare our experimental results with earlier findings of theoretical and experimental nature. Fig. 5.33 serves this purpose. Here we have plotted a JONSWAP (Hasselmann et al., 1973) for a 52 km fetch with a result obtained on 13th March, 1981 using the multifrequency radar. We can see from this comparison that there is a good general agreement.

Here it is important to note, as pointed out earlier, that no attempts have been made to convert our normalized spectral intensity to a quantity that provides direct information about wave height. The relation between our normalized spectral intensity and wave height squared per hertz is not likely to be a linear one, neither is it likely to be invariant for varying conditions of

128 *Sea clutter background: characterization of ocean surface*

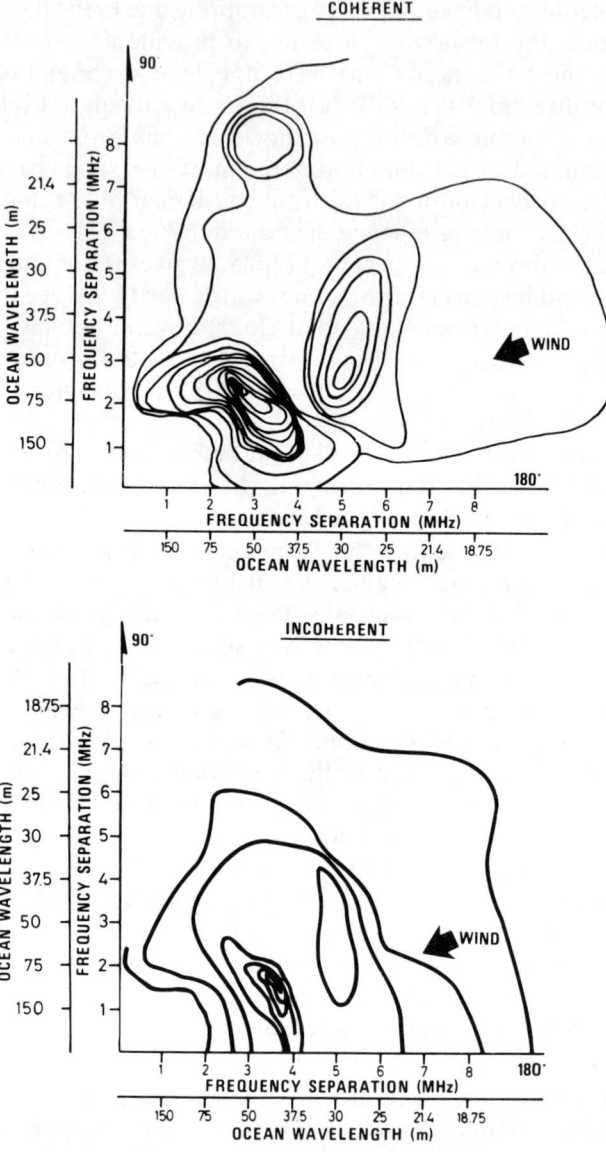

Fig. 5.32 *Two-dimensional ocean irregularity spectrum (isolines of $\Phi(\Delta F)$) when the wave height has increased to 2·5 m peak-to-peak and the wind has turned towards east*

whitecapping (splash), foam formations, etc. Therefore, all that Fig. 5.33 can tell us is that the spectral wave intensity distributions observed by Hasselmann *et al.* are similar to those observed by our multifrequency radar system.

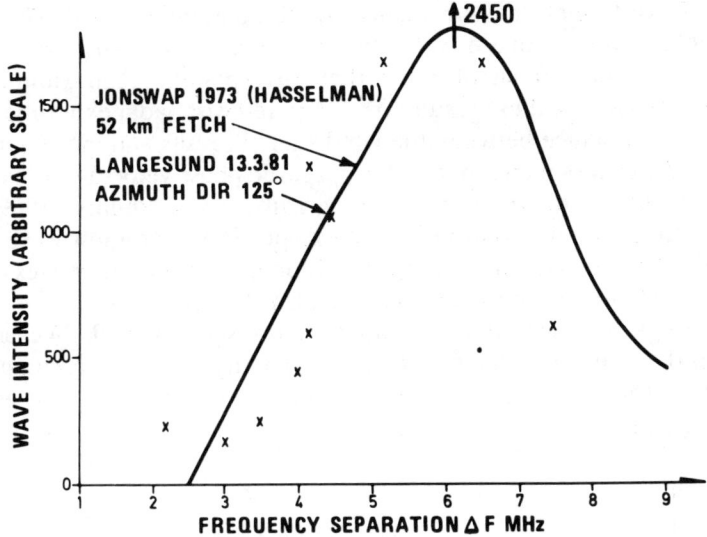

Fig. 5.33 *Typical result from the multifrequency radar experiment is compared with the 'classical' JONSWAP experiment (Langesundsfjorden 12th March, 1981, 12.15)*

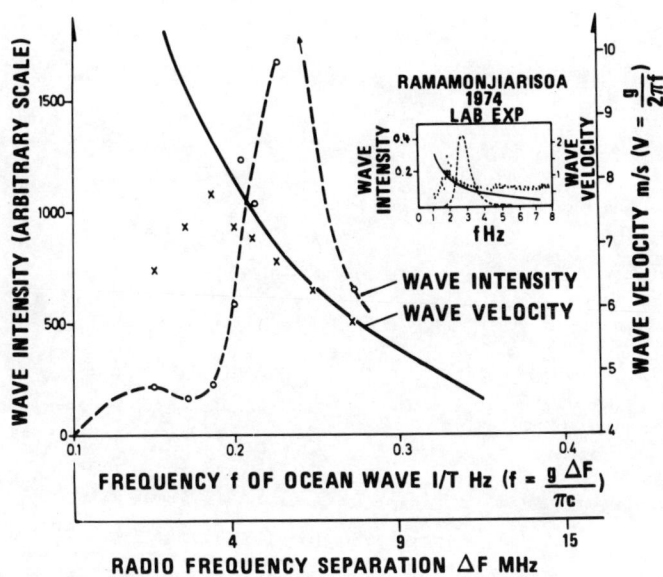

Fig. 5.34 *Wind-tunnel experiments of Ramamonjiarisoa (1974) are compared with the results from multifrequency radar experiments (Langesundsfjorden, 12th March, 1981: Gjessing and Hjelmstad, 1986)*

Wind wave-tunnel measurements by Ramamonjiarisoa (1975) show a remarkable drop in phase velocity of ocean waves with wave number (oscillation frequency), smaller than that corresponding to maximum spectral intensity. In Fig. 5.34 these results are compared with radar experiments. Note the close resemblance between the wind-tunnel results and those obtained by the radar. Near the spectral peak, the measured phase velocities lie close to the dispersion relation for freely travelling gravity waves. Phillips (1969, p. 157) attributes the marked deviation from the dispersion line for low wave numbers to disturbances associated with groups of the more energetic waves travelling with the appropriate group velocity (see also Phillips, 1957).

This brings the section on ocean signatures to an end. We shall now approach the more complex problem of providing matched illumination for a target in the form of a rigid ship against an ocean background. This will be the topic of Chapter 6.

Chapter 6

Ship targets against a sea clutter background

The task we are now facing is to form the basis for an optimum estimation of target parameters to provide the computer system with a protocol based on which a strategy for a general adaption procedure can be developed.

We shall confine ourselves to a discussion of ship targets against an ocean background. Furthermore, we shall characterize the ship target against the sea background in three signature domains: space, motion pattern, and space/time coherence.

6.1 Spatial signature (wave-number matching)

In the preceding sections, we have shown on the basis of general physical principles that the correlation properties of the radio waves scattered back from an object at distance z_0 are given directly as the Fourier transform of the delay function characterizing the object. Thus, if the object is a rectangular one, with length Δz at a distance z_0, then the envelope of the autocorrelation function in the frequency domain of the returned radio waves is simply a $(\sin x)/x$ relationship, the width of which is $c/2\Delta z$ and under which envelope there are sinusoidal oscillations, with period $c/2z_0$.

We shall now consider the spatial signatures of ships. In order to calculate a detailed signature as it appears at the radar station, we shall have to know how the scattering cross-section is distributed along the target. In the practical case, this will not be a smooth function. From experience we know that the 'bulk' scattering cross-section is made up of a finite number of discrete diffuse scattering centres which physically manifest themselves as imperfect corner reflectors. The delay function of a ship target (the Fourier transform of which is the radar signature) will then probably consist of a limited number of unequally spaced δ functions. From the conventional diagrams of scattering cross-section, plotted versus the azimuthal aspect angle, we can, to some degree, deduce the delay functions (see Fig. 2.19).

In Fig. 2.7 a set of target delay functions are shown together with their resulting radar signatures. Curve B of Fig. 6.1 shows the multifrequency radar signature of a 100 m ship characterized by an even distribution of scattering centres, whereas curve A shows the signature of one consisting of two dominating scatterers 100 m apart.

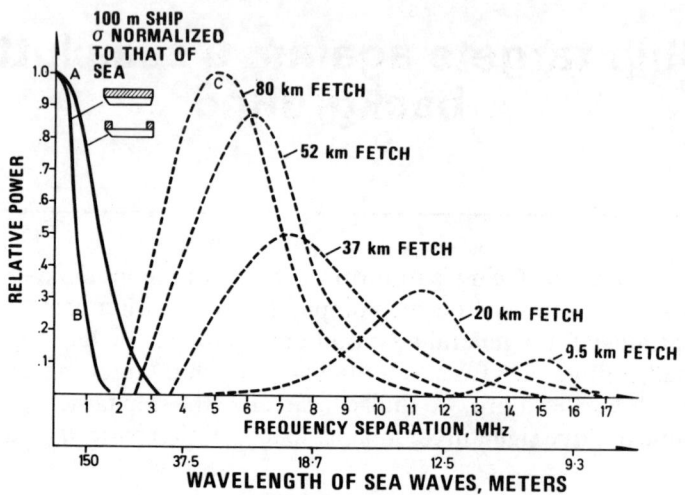

Fig. 6.1 *Signatures of ships compared with sea surface signatures*
Note that the sea calculations are based on the JONSWAP experiments and that the relative power scale is arbitrary (Gjessing, 1981a © IEEE).

As emphasized in Chapter 2 on basic principles for wave-number matching, there are great differences in the K domain between periodic and non-periodic delay functions. As a result of this, the target contrast against a sea clutter background can be made very large when matched illumination is employed.

This is shown in Fig. 6.1 and is even more pronounced in Fig. 6.2, where experimental sea clutter is compared with a 'Gaussian ship'. Note that the characterizing delay function of the ship is assumed to be Gaussian in all azimuth directions, the length of the ship being 100 m and the width 25 m. Note that in Fig. 6.2 the ship is heading in the direction of the dominating ocean waves. If, however, the heading of the ship is changed by 90°, the ship signature will overlap considerably that of the sea surface.

Note that an experimental ship signature has already been illustrated in Fig. 3.8.

This, naturally, is well known from experience. It is very difficult to detect a small ship (length comparable with the dominant sea wave length) against a high sea state background.

In the next section a remedy making use of the temporal properties of the signal return will be discussed.

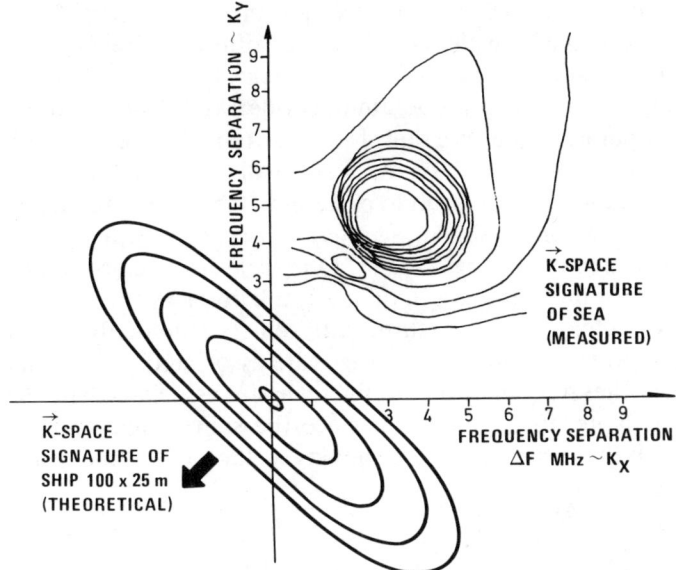

Fig. 6.2 *K-space representation of the sea surface (isolines of spectral intensity $\Phi(K)$) as obtained from experiments is compared with the K-space signature of a ship with the K-space signature of a ship with Gaussian distribution of the scattering elements*

6.2 Temporal (Doppler) signatures

In the previous section we saw that, knowing the delay function of the target, an illumination function can be structured (composition of wavelengths) to obtain maximum signal/noise ratio for the target of interest, and also to obtain a target identification capability. We shall now introduce another aspect; namely, that of temporal (Doppler) filtering. We illuminate the target and its background with a set of electromagnetic waves having different matched wavelengths and we observe the Doppler shift and Doppler broadening associated with each of the back-scattered radar waves. For the purpose of specificity, let us base the following discussion on Fig. 3.3 showing a particular experimental multifrequency radar system.

In this system, the radar transmitter consists of six mutually correlated frequency generators. The receiver, on the other hand, has six very narrowbanded receiving amplifiers, which are synchronized with the transmitters. The frequencies of the six transmitters as well as that of the six receivers, are chosen to provide 15 different frequency separations. We now measure the Doppler shift and the Doppler broadening associated with each of these 15 frequency pairs simultaneously. This means that we can look at 15 different scale sizes, both in the target and at the sea surface, and we can measure the rate at which these scales move one half beat frequency wavelength (one half wavelength along the difference frequency scale).

Note that if the number of independent frequency lines available is organized in this manner, we obtain only one $E(\omega)E^*(\omega + \Delta\omega)$ product per frequency separation $\Delta\omega$. We therefore have to make use of time averaging in order to obtain estimates which are statistically meaningful.

Introducing the concept 'motion pattern matching', the reader is referred to Fig. 6.3. Here we have shown a target in the form of a ship and we have indicated two dominant target scales. One scale size is constituted by the bow and the stern of the ship (length of ship), the other by the ship's funnels. We choose two frequency pairs (three different but mutually correlated radio frequencies) to match one pair to the scale associated with the length of the ship, and another frequency pair to the scale size associated with the spacing of the funnels. Note, in passing, that since the ship is rigid and in translatory motion, both these scale sizes are moving at the same velocity V_T.

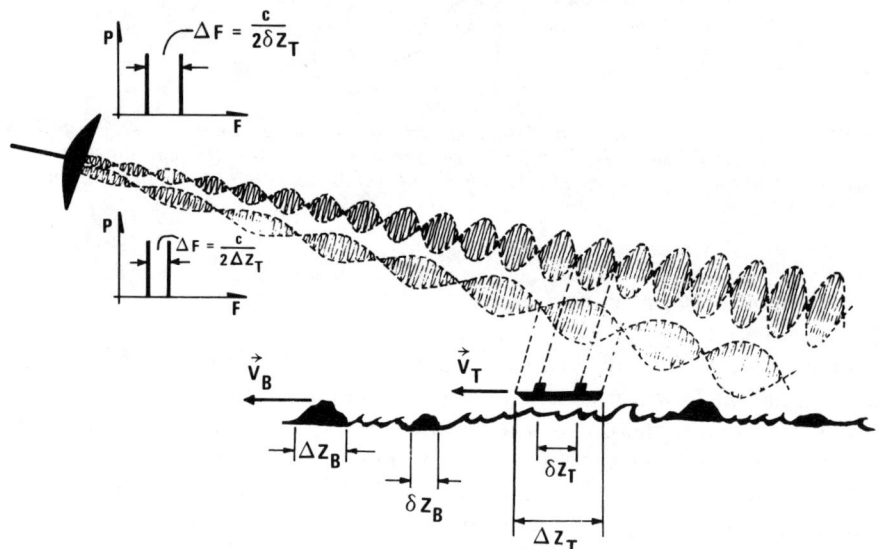

Fig. 6.3 *A simple three-frequency adaptive radar system*
One frequency pair is coupled to the total length of the ship target whereas the second pair is coupled to another set of scattering centres. Note that the echo signal from the target competes with that from the sea surface only where the irregularity scales of the ocean are the same as those of the target (Gjessing, 1981a © IEEE)

We now want information about the Doppler shift to which this frequency ΔF is subjected. We base this calculation on the basic Doppler equation, the way it was introduced in Section 2.4

$$f = \frac{1}{2\pi} \mathbf{K} \cdot \mathbf{V}$$

Here $K = 2\pi/L$, where L is the scale size to which the frequency pair ΔF is matched. The Doppler shift associated with the difference frequency ΔF is therefore

$$f = \frac{1}{2\pi} K \cdot V = \frac{1}{2\pi} \frac{2\pi}{L} V \cos \phi$$

$$= \frac{2\Delta F}{c} V \cos \phi \qquad (6.1)$$

where ϕ is the angle between the wave vector K and the velocity vector V.

Hence, if we illuminate a moving object with two electromagnetic waves with frequency spacing ΔF (coupled to scale size $L = c/2\Delta F$), this frequency ΔF is subjected to a Doppler shift f which is proportional to ΔF and to the velocity V of the object. The power associated with the Doppler frequency f is the same as that associated with the freuqency ΔF and expressed by eqn. 2.4.

Thus, knowing the shape of the target (see Fig. 2.7) and its velocity, the Doppler spectrum can be calculated on the basis of eqns. 2.4 and 6.1.

Then let us again consider the Doppler spectra resulting from a backscattering from the sea surface. We shall then bear in mind that gravity waves are dispersive. Phase velocity V is given by

$$V = \sqrt{\frac{gL}{2\pi}} \qquad (6.2)$$

Hence

$$f = \frac{2\Delta F}{c} \sqrt{\frac{gL}{2\pi}} \qquad (6.3)$$

In order to couple constructively to the ocean wave of wave length L a difference frequency ΔF should be used where ΔF is given by

$$\Delta F = \frac{c}{2L}$$

Hence

$$f = \sqrt{\frac{g\Delta F}{\pi c}} \qquad (6.4)$$

For a rigid object in motion, all scales within the object move at the same velocity for translatory motions. When motion also involves a change in pitch, roll and yaw, there is a linear deterministic relationship between the motion of the various scales. The Doppler shift caused by a ship is, as we have already noted, therefore given by

$$f = \frac{2\Delta F}{c} V_{SHIP} \qquad (6.5)$$

Thus, the Doppler spectrum from the various difference frequencies ΔF coupling to various irregularity scales within the rigid body is centred round a Doppler frequency which is proportional to ΔF.

As an example, Fig. 6.4 shows the result of simple calculations (eqns. 6.4 and 6.5, respectively) based on a situation where a ship (30 m long) moves with the same velocity as that of an ocean wave with the same dominant wave length as the length of the ship (30 m). From a radar detection point of view, this represents the greatest challenge (see Fig. 6.1).

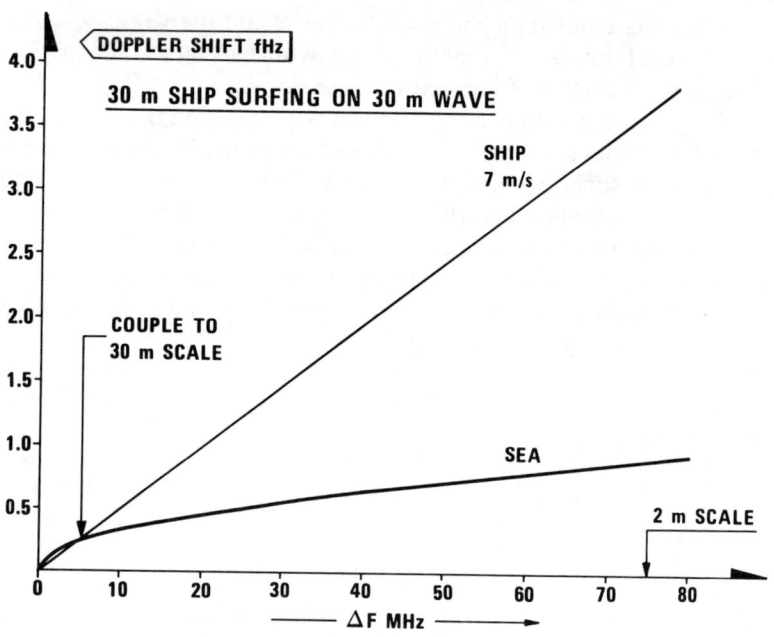

Fig. 6.4 *Illuminating the ship with a set of matched frequencies, the target Doppler signature differs substantially from that of the sea background owing to the dispersive behaviour of ocean waves. Note that we have assumed the worst case with a 30 m ship 'surfing' on a 30 m dominant wave*

Since we know the relationship between ocean wave velocity and ocean wave length, we can construct simple step filters to exclude the contribution from sea clutter. The characteristics of such filters are illustrated in Fig. 6.5. No matter what the speed of the ship, there is only one ocean wave length that can match the speed of the ship and thus contribute to the target signal. In our particular radar shown earlier, we therefore have another 14 extra values of ΔF where the sea contribution is suppressed. In order to ensure a thorough physical understanding of this very important property of the matched

Ship targets against a sea clutter background 137

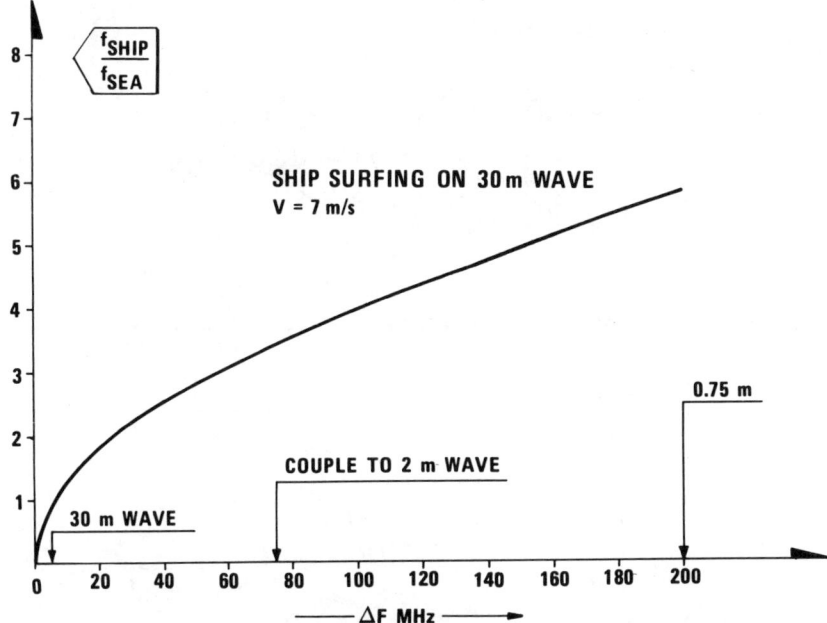

Fig. 6.5 *Doppler shift from a ship relative to that from the competing sea surface is plotted to the basis of target matching frequency ΔF*
Note that when a 30 m ship (speed 7 m/s) is 'surfing' on a dominant 30 m wave $f_{SHIP}/f_{SEA} = 1$ when ΔF corresponds to a 30 m wave length. At other values of ΔF the Doppler shifts are different and the contribution from the sea can be filtered out

illumination concept in relation to surface ships against a high sea state background, Fig. 6.6 is presented.

By the use of six correlated and highly coherent microwaves, 15 different frequencies are produced. These are chosen to lie between 5 and 75 MHz so as to be matched to scales within the target ranging from 30 m (total length of target) to a small scale of 2 m.

The lower left-hand curve giving target shape shows that there is considerable overlap with the sea surface signature when a 30 m ship is surfing on a 30 m dominant ocean wave. Inspecting the right-hand Doppler diagram, however, we see that due to the dispersiveness of ocean waves, there is a very marked difference between ship and sea signatures. This is further illustrated in Fig. 6.7, showing the effect of changing the target velocity from 10 knots to 20 knots.

In concluding this section on motion pattern signatures, Fig. 6.8 is presented as an experimental verification of the theoretical results of Fig. 6.7.

138 *Ship targets against a sea clutter background*

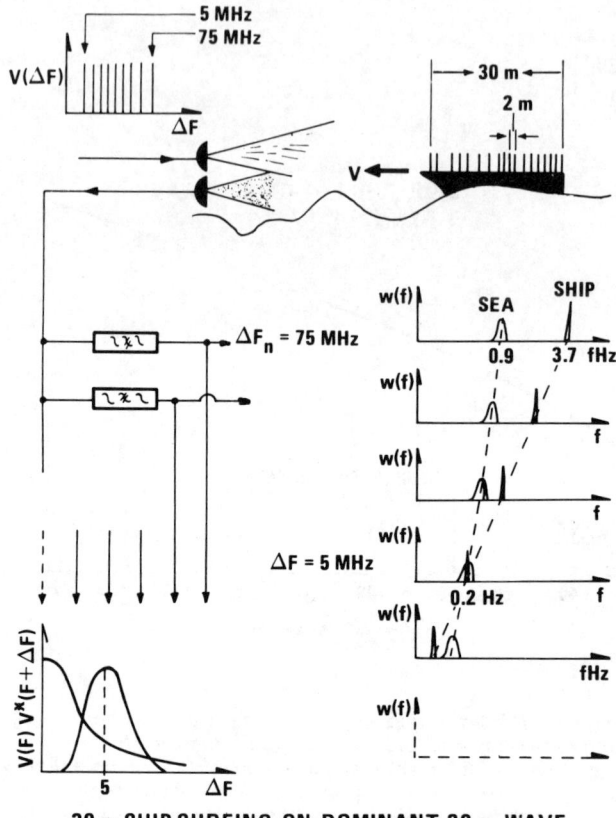

Fig. 6.6 *30 m ship surfing on a dominant 30 m ocean wave*
A set of radar frequencies ΔF is matched to the target. The lower left curve giving the target shape shows a considerable overlapping of target and clutter signatures. This is, however, not the case in the Doppler domain

6.3 Space/time (mutual) coherence: target detection enhancement based on the measurement of rigidity

This topic was first introduced in Section 2.5. Fig. 2.37 illustrates the basic concepts with particular emphasis on the target identification potentials of rigidity information through the computation of mutual coherence.

In this section we shall not focus on the non-co-operative IFF concept (identification) but rather concentrate on enhancement of the target/clutter ratio.

As we have already emphasized, the target/clutter ratio can be improved through the multifrequency matched illumination method in several ways:

1 Matching the illumination (shape and size) to the target, reduce the coupling to the sea surface and hence its influence. This is emphasized in Figs. 6.2 and 6.3.

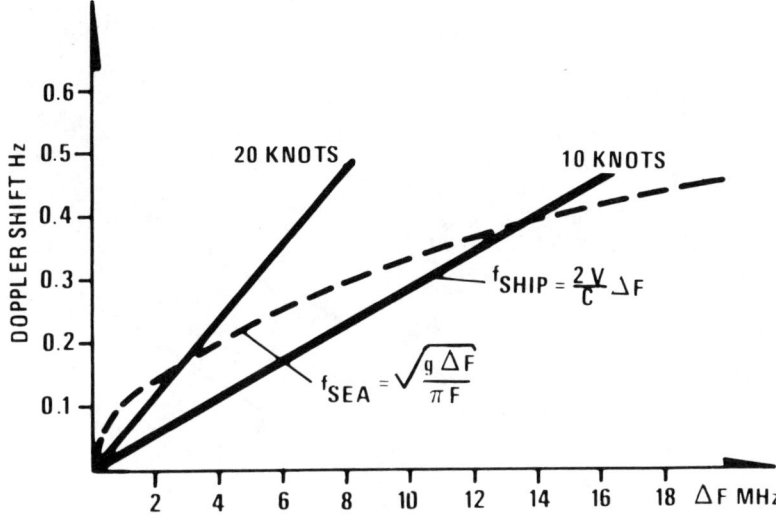

Fig. 6.7 When the target is illuminated with a set of matched frequencies, there will only be one frequency which is subjected to the same Doppler shift from the target as that from the sea surface due to the square root dispersive relationship of ocean waves. Based on this property, it is possible to reduce the effect of sea clutter substantially

2 Ocean waves are dispersive, the motion of a rigid ship is not. This means, as emphasized in Figs. 6.4 to 6.7, that we can make use of the fact that the Doppler shift produced by the ship is proportional to frequency separation ΔF whereas that produced by the dispersive sea is proportional to $\sqrt{\Delta F}$. The sharper we make the sea clutter filter, the larger is the target/clutter ratio.

3 The ocean waves are compressible, the scatterers constituting the target are integral parts of a semi-rigid steel body. Hence, the wave-number components (scales), the integral of which is the target scattering cross-section, tend to move, and also vibrate, in unison, whereas this is not the case for ocean scatterers which ride on a compressible wave pattern.

We shall now briefly consider these mutual coherency effects and give some experimental examples from a comprehensive investigation which currently is in progress at the author's organization (Dittel, Gjessing, and Hjelmstad, 1985).

We illuminate the target with $n = 6$ frequencies, thus obtaining $N = 15$ values of ΔK

$$N = \frac{n(n-1)}{2}$$

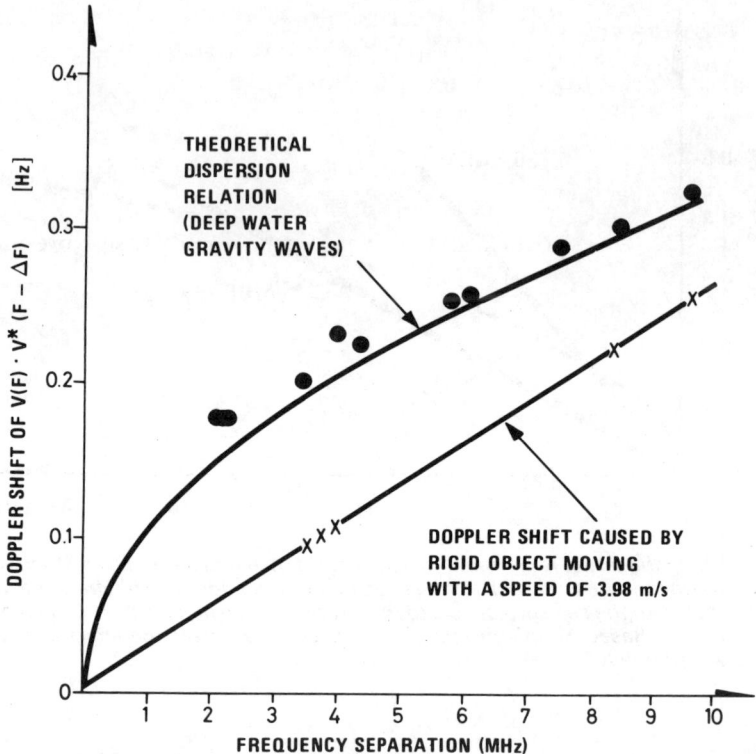

Fig. 6.8 *Doppler shift as a function of frequency separation for a rigid target and for sea clutter, respectively*
Note that deep sea gravity waves are dispersive: there is a square root relationship between phase velocity and wave length. A rigid body moving at constant velocity gives a Doppler shift that is proportional to the frequency separation and the velocity of the object

When we compute the mutual coherence we form the cross-spectrum between the quantities $A_1(\Delta F_1, t)$ and $A_m(\Delta F_m, t)$ where the A-functions are defined as follows (see Section 2.5)

$$A(\Delta F, t) = V(F, t)\, V^* \{(F + \Delta F), t\}$$

Hence, if we have six frequencies ($n = 6$) and hence 15 beat frequencies ($N = 15$), we shall have

$$m = \frac{N(N-1)}{2} = 105$$

cross-spectra.

This means that we shall have an ensemble of 105 estimates of rigidity.

Obviously, for a rigid or semi-rigid body with a finite number of scattering elements (scattering centres) it is not possible to obtain N independent ΔF values from n values of F. By the same token, we cannot expect to find m independent cross-spectra.

Before the work in progress, referenced above, has been brought to a conclusion, we shall not focus on the question of 'effective' ensemble but confine ourselves to a presentation of illustrative examples.

First consider the sea surface and in particular the data sets presented in Figs. 5.29 to 5.34.

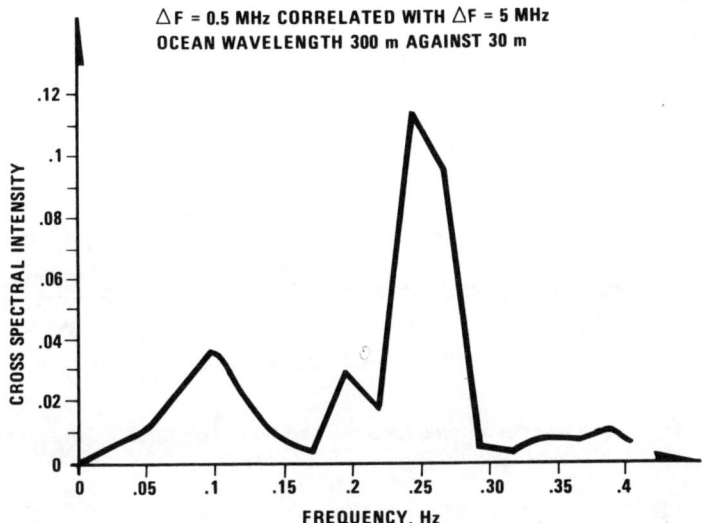

Fig. 6.9 *Cross-spectrum estimates for an ocean wave length of 300 m against one 30 m long (Jacobsen, 1985)*

As illustrated in Fig. 2.36, before we can derive the cross-spectrum between $A_1(\Delta F_1, t)$ and $A_2(\Delta F_2, t)$ we shall have to stretch the time axis so as to make the spectrum (autospectrum) of A_1 coincide on the frequency axis with that of A_2. The consequence of this statement is illustrated in Fig. 6.6.

In addition to forming the basis for target/clutter enhancement as emphasized above, it is also of interest from a fluid dynamics point of view to investigate the cross-spectrum properties of ocean waves. For example, through suitable choice of ΔF we couple to a 300 m wave ($\Delta F = 0.5$ MHz) and at the same time to a 30 m wave ($\Delta F = 5$ MHz) and by means of cross-spectrum computations we investigate the degree to which the various temporal spectral components correlate. Are we likely to find some degree of cross-coherency, e.g. from the assumption that the 30 m wave component 'rides on' the 300 m wave. Could we argue that as a result of the normally occurring cusp on the

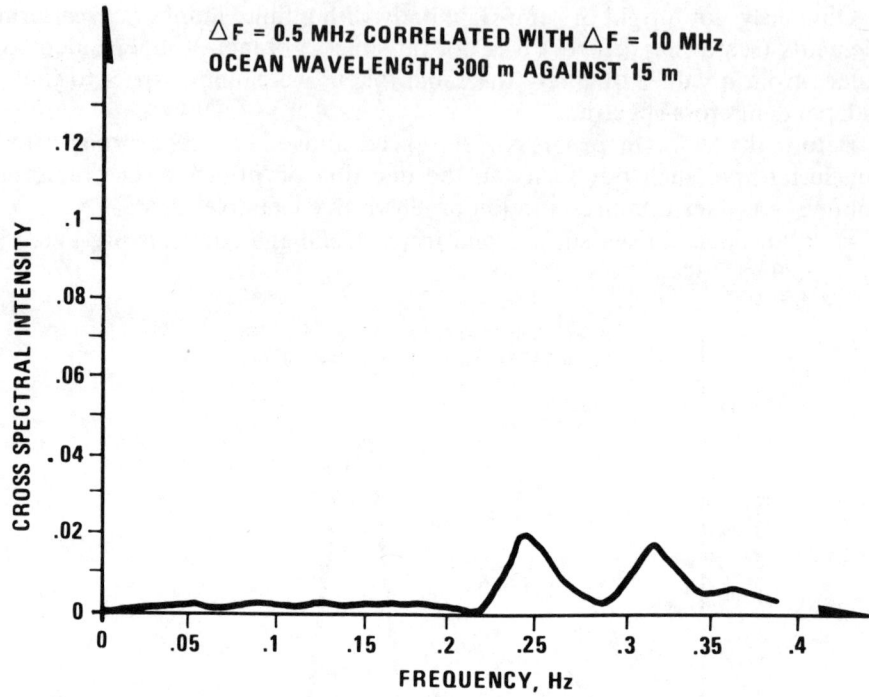

Fig. 6.10 *Cross-spectrum estimates for an ocean wave length of 300 m against one 15 m long (Jacobsen, 1985)*

ocean wave one should expect to have higher order harmonic components that are in fact linked to the long wave and this would be expected to give a certain degree of cross-spectrum correlation. Such questions are addressed by a research group affiliated with the author at Tromsø University. Examples from the investigation (Jacobsen, 1985) are given in Figs. 6.9 and 6.10.

Note that in order to provide anything but comparative results, the normalization factors employed will have to be considered very carefully. The experimental results are therefore to be regarded as examples from which general conclusions about the dynamics of ocean waves should not be drawn. However, when carrying out the same general analysis (forming 105 cross-spectra) with a rigid ship in the beam, a cross-spectrum density (mutual coherency) which is more than two orders of magnitude larger than for the sea surface is obtained. This indicates that even with the minimum of two sets of beat frequencies (three values of F), thus producing one cross-spectrum, a considerable clutter reduction is to be expected. With an ensemble of 105, some additional clutter rejection is to be expected (Dittel, Gjessing, and Hjelmstad 1985).

Before bringing this section on space/time mutual coherence to an end, a 3-dimensional 'association plot' for ocean gravity waves will be presented as a crude illustration of the potential of the target adpative multifrequency technique. Here, the vertical axis gives cross-spectral intensity whereas ΔF_1 and ΔF_2 are plotted orthogonally in the horizontal plane, see Fig. 6.11.

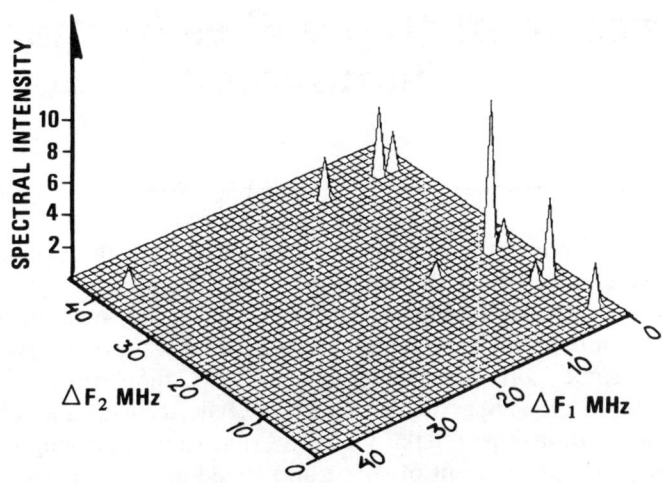

Fig. 6.11 *'Association plot' for ocean gravity waves*
150 m ocean waves and 30 m waves have a comparatively high degree of mutual association
Ocean waves in the order 150–100 m are more associated with short waves than with long ones
Long waves tend to be poorly associated with each other
(Jacobsen, 1985)

Again note that these illustrations should be regarded as samples from which no general conclusion should be drawn. One conceivable reason for the lack of mutual coherence between long waves could be that the wave energy (wave height) in the particular data set was much lower for long ocean wave lengths than for short ones.

Chapter 7

Detection of ship wakes by matched illumination

As indicated already in the introduction, there are many targets that produce a footprint which gives much information about the characteristics of the target itself. One such example is the wake produced on the sea surface by a ship. Such wave phenomena have been the subject of comprehensive studies for many years since Lord Kelvin, as early as 1883, first formulated the problem mathematically. Specifically, Lord Kelvin derived the characteristic geometrical relationships for the V-shaped ship waves, proving that the apex angle is 39° and independent of shape and speed of the ship.

This, obviously, is an ideal situation for a matched illumination radar: knowing the geometry of the wave pattern, the challenge involves the generation of a radar wave pattern that matches the wave pattern of the sea surface. Providing such a matched illumination results in a dramatic enhancement of the contribution to the received signal from V-shaped wakes whose apex angle is 39° at the expense of contributions from other wave patterns.

Fig. 7.1 shows an aerial photograph (from Newman) of a typical ship wave pattern, whereas Fig. 7.2 illustrates the same phenomenon as observed by the synthetic aperture radar of SEASAT.

The simple theory explaining the wave phenomena observed in Figs. 7.1 and 7.2 is based on the following assumptions:

1 There is one dominating interface only, air/sea. The density of the deep sea is constant with depth.
2 In the reference system of the ship, the wave pattern is stationary such that the wave frequency is zero.
3 There is one wave generator only, namely that constituted by the ship. The energy radiates from the ship.

A closer examination of SAR images, however, reveals additional wave patterns with apex angles much smaller than the 39° predicted by Lord Kelvin. These, presumably, are due to the existence of internal waves set up by a

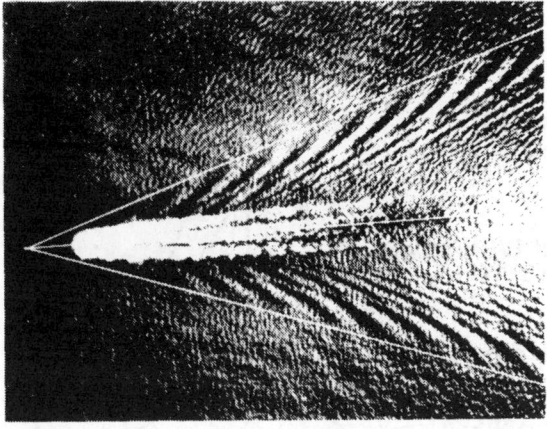

Fig. 7.1 *Ship wakes observed from above. Note that the lines marked make 19·5 degrees, relative to the direction of the ship*

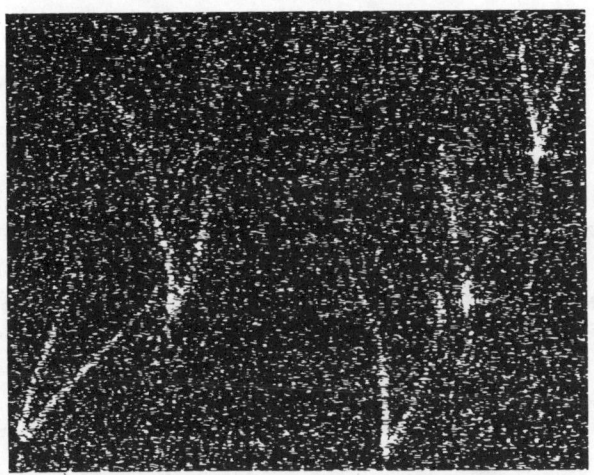

Fig. 7.2 *SAR-image processed by Royal Aircraft Establishment, England*

shallow thermocline (marked step in the density profile) which is distributed by the action of the ship (see eqns. 5.6 and 5.7).

An idealized wave pattern based on basic first-order physics to be presented shortly is illustrated in Fig. 7.3. For details regarding this subject the reader is referred to Shuchman *et al.* (1983).

146   Detection of ship wakes by matched illumination

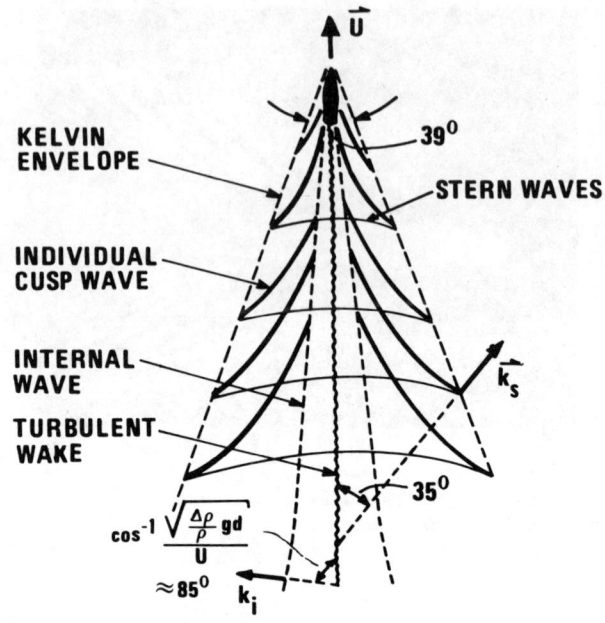

Fig. 7.3 *Many wave phenomena contribute to the ship wake signature: surface gravity waves, internal waves and turbulent wakes (Shuchman et al., 1983)*

On the basis of fundamental work presented earlier (in particular by Dysthe and Gjevik) we shall now consider a simplified treatment aiming at forming the basis for a matched illumination scheme in relation to the two-wave phenomena generated by a moving ship: dispersive surface waves and non-dispersive internal waves.

7.1 Directional properties of gravity waves generated by ships: detection of ships by matching the illumination to the ship's wake

The purpose of this section is to show from basic physical principles that a ship produces a very definite wake pattern that forms the basis for detection.

We shall also design a 'matched illumination, matched reception' system that gives maximum detection probability on the basis of knowledge about the wake signature.

The dispersion relation for a surface gravity wave in shallow water is given by

$$\omega^2 = kg \tanh(kd) \tag{7.1}$$

where

> ω = wave frequency
> k = wave number $2\pi/L$ where L is ocean wave length
> d = water depth
> g = grvitational constant

For shallow water

$$\tanh(kd) \simeq kd$$

Hence

$$\omega^2 = gdk^2 \tag{7.2}$$

The phase velocity of the wave is therefore

$$v = \frac{\omega}{k} = \sqrt{gd} \tag{7.3}$$

i.e. no dispersion.

For deep water $\tanh(kd) \simeq 1$ and we have

$$\omega^2 = gk$$

The phase velocity is then

$$v = \frac{\omega}{k} = \sqrt{\frac{g}{k}} \tag{7.4}$$

and the group velocity is given by

$$v_g = \frac{d\omega}{dk} = \frac{1}{2}\sqrt{\frac{g}{k}} \tag{7.5}$$

In the reference system of the ship, the ocean has a velocity $-U$ as indicated in Fig. 7.4

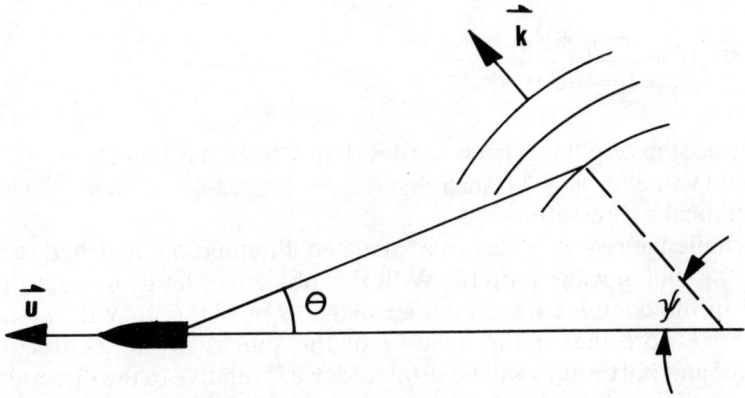

Fig. 7.4 *Ship moving at velocity **U** produces a wave pattern with a direction ψ relative to the direction of the ship*

The frequency ω of the surface wave will therefore as a result of the Doppler shift, be $-\mathbf{k}\cdot\mathbf{U}$, i.e.

$$\omega = \sqrt{gk} - \mathbf{k}\cdot\mathbf{U}$$

(see Dysthe, 1980).

For the ship wave $\omega = 0$, i.e.

$$\sqrt{gk} = \mathbf{k}\cdot\mathbf{U} \qquad (7.6)$$

Referring now to Fig. 7.4 we see that the angle between $\mathbf{k}$ and $\mathbf{U}$ is ψ.

Hence

$$\sqrt{gk} = KU\cos\psi$$

i.e.

$$\frac{g}{k} = U\cos\psi \qquad (7.7)$$

However, we have already established an expression for the group velocity of the surface wave

$$v_g = \frac{1}{2}\sqrt{\frac{g}{k}}$$

Accordingly

$$v_g = \tfrac{1}{2}U\cos\psi \qquad (7.8)$$

From the geometry of Fig. 7.4 we find

$$\tan\theta = \frac{\cos\psi \sin\psi}{2 - \cos^2\psi}$$

i.e.

$$\tan\theta = \frac{\cos\psi \sin\psi}{1 + \sin^2\psi} \qquad (7.9)$$

This expression originates from Lord Kelvin (1883). This function will have its maximum value for $\psi = 39°$ such that $\theta_{max} = 19\cdot5°$. Figs. 7.1 and 7.2 verify this mathematical expression.

The challenge now is to design a 'matched illumination, matched reception' system for such a wake pattern. With this objective, let us present the wake pattern in the complex wave-number plane as in Fig. 6.2. With reference to Fig. 7.5 we note that if the heading of the ship is in the Y-direction, the starboard and port wakes will be displaced $\pm 35°$ relative to the direction of the ship, as indicated in Fig. 7.5. The stern wave will be in the $-Y$-direction.

With reference to the arguments presented, for example, in connection with

MACHED ILLUMINATION FOR DETECTION OF SURFACE SHIP-WAKES

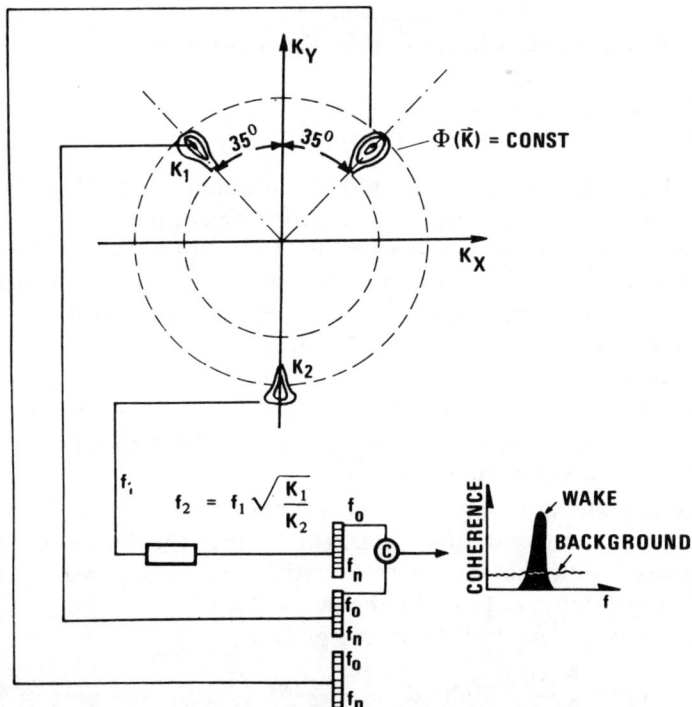

Fig. 7.5 *A ship produces a well-defined wave pattern which is independent of its velocity and displaced ±35° relative to the direction of the ship*

Fig. 5.8, we see that two antennas displaced by 70° employing the same multifrequency illumination matched to the wave length of the wake would provide the desired detection enhancement (see Sections 6.1, 6.2 and 6.3 above).

This concludes the simple derivations based on first-order physical concepts in relation to surface gravity waves.

We shall now, with reference to Sections 5.1 and 5.2, estimate the effect of internal waves stimulated by surface disturbances on the wake pattern.

7.2 Directional properties of internal waves generated by a disturbance of a surface ship

As mentioned in the introduction to this section on ship wakes, there is experimental evidence suggesting that in addition to the generation of the 39° V-shaped surface gravity waves, a deep keeled ship disturbs a shallow thermocline, thus setting up internal waves. In general there are several different wave patterns resulting from the motion of a displacement ship:

1 The Kelvin or bow wave envelope
2 The cusp waves which form the Kelvin envelope
3 The stern waves
4 The turbulent wave
5 The internal waves

For details the reader is referred to Shuchman et al. (1983).

We shall now study the general properties of internal waves. Note that the objective is to give first-order results so as to form the basis for wake detection by matched illumination rather than presenting comprehensive expressions. The following simple approach is based on earlier works by Dysthe (1980), Gjevik (1983), and Gjevik and Martinsen (1978).

Consider, as in Section 5.1, a two layered ocean. The top layer has density ρ and extends to a depth d. The bottom layer is infinitely deep and has density $\rho + \Delta\rho$. The thermocline at depth is disturbed by the keel of the ship moving at velocity U as depicted in Fig. 7.6.

In exactly the same way as for the air/sea interface where the restoring force is a result of the gravitational constant g (since the density of air is very small compared with that of the sea) in the case of the two layered ocean, the restoring force is reduced by the factor $\Delta\rho/\rho$.

The wave equation is therefore given by

$$\omega^2 = g\frac{\Delta\rho}{\rho}k \tanh kd \qquad (7.10)$$

the group velocity is then

$$v_g = \frac{d\omega}{dk} = \begin{cases} \sqrt{\dfrac{\Delta\rho}{\rho}\dfrac{g}{k}} & \text{when } kd \gg 1 \\ \sqrt{\dfrac{\Delta\rho}{\rho}gd} & \text{when } kd \ll 1 \end{cases} \qquad (7.11)$$

FOOTPRINT FROM SUB-SURFACE OBJECTS

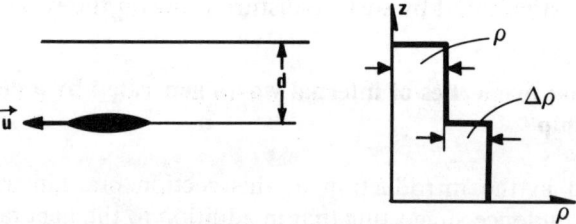

Fig. 7.6   *Thermocline at depth d is disturbed by the keel of a ship moving at velocity **U***

The group velocity is a maximum for large wave lengths (or shallow thermoclines), and so is the phase velocity (Fig. 7.7).

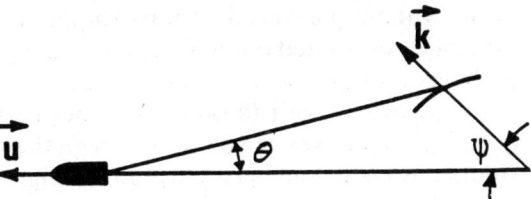

Fig. 7.7 Symbol definition

$$\tan \theta = \frac{v_g \sin \psi}{u - v_g \cos \psi} \quad \text{(i)}$$

$$\frac{\omega}{k} = u \cos \psi \quad \text{(ii)}$$

(7.12)

ψ is a minimum when $kd \ll 1$

$\psi = \cos^{-1} \gamma$

when the ratio γ is given by

$$\gamma = \frac{\sqrt{[(\Delta \rho / \rho) gd]}}{\bar{u}} \quad (7.13)$$

Note that γ is the ratio of the group velocity of the internal wave and the velocity $\bar{u}$ of the disturbing object. The inverse of this ratio ($\bar{u}/\sqrt{\{[\Delta \rho / \rho] gd\}}$) is known as the internal Froude number.

This is the condition for minimum angle ψ and hence for maximum θ. The maximum value of θ is then given by

$$\theta_{max} = \tan^{-1} \frac{\gamma}{\sqrt{(1 - \gamma^2)}} \quad (7.14)$$

As an example, consider a typical thermocline at depth $d = 100$ m and $\Delta \rho / \rho = 10^{-3}$. The group velocity then becomes

$$\sqrt{[(\Delta \rho / \rho) gd]} \sim 1 \text{ m/s}$$

and the angle θ is very small (for details, the reader is referred to Gjevik, 1983; Dysthe, 1980).

As an illustration in regard to typical directional internal wave patterns, Fig. 7.8, which is taken from Gjevik (1983), is presented.

As a summing up of this section on the directional properties of internal waves, Fig. 7.9 is presented. Here, as an example, typical values of thermocline depth $d = 100$ m and density gradient $\Delta\rho/\rho = 10^{-3}$ are used as the basis for the calculations.

So far we have considered the directional properties of internal waves in rather broad terms and without the consideration of amplitudes or coupling to surface wave phenomena. As pointed out in Chapter 5, however, see Fig. 5.2, the internal waves have a pronounced effect on surface waves moving at velocities which are comparable with those of the internal waves. Through these interactions, the internal waves can be observed on the sea surface when a radar having a suitable wave length is employed (see Fig. 5.3).

These phenomena will now be considered (see also Apel *et al.*, 1975, and Apel and Gonzales, 1983).

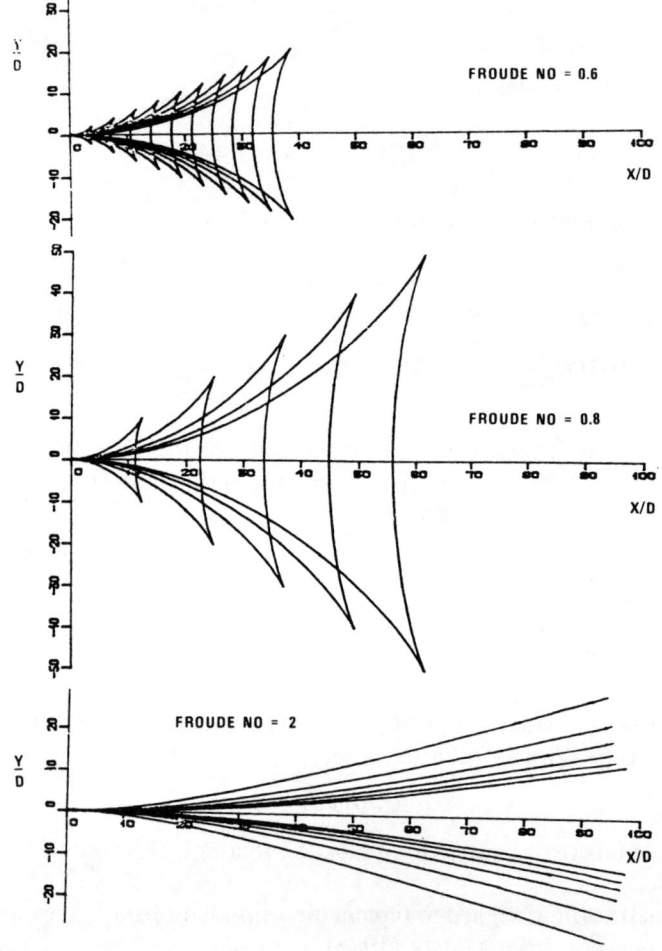

Fig. 7.8 *Directional patterns of internal waves for different values of the Froude number (from Gjevik, 1983)*

Detection of ship wakes by matched illumination

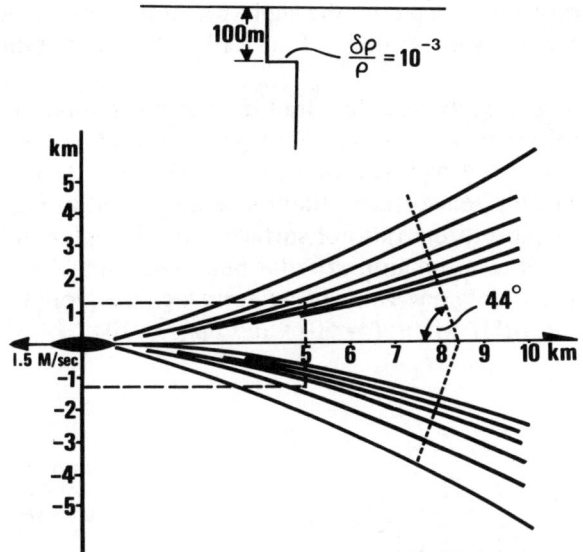

Fig. 7.9 *If a ship moving at 3 knots disturbs the thermocline, a characteristic internal wave pattern is generated. Knowledge about this forms the basis for matched illumination*

7.3 Coupling mechanisms between internal waves generated by a ship and surface irregularities

In Chapter 5 (see eqn. 5.10) it was shown that internal waves have a dramatic influence on surface waves when their group velocities match.

Since surface gravity waves are dispersive ($v = \frac{1}{2}\sqrt{(g/k)}$) whereas internal waves on a shallow thermocline are non-dispersive ($v = \sqrt{\{(\delta\rho/\rho)\,gd\}}$), there is one particular ocean surface wave length L which is affected for a given $(\delta\rho/\rho)d$, namely

$$\Delta F = \frac{c}{2L} = \frac{c}{16\pi(\delta\rho/\rho)\,d}$$

The footprint on the sea surface of the internal wave moves with a velocity $\sqrt{\{(\delta\rho/\rho)\,gd\}}$ giving a Doppler shift which is linearly dependent on ΔF

$$f_i = \frac{2\Delta F}{c}\sqrt{\left(\frac{\delta\rho}{\rho}gd\right)} \tag{7.15}$$

whereas the surface gravity waves move with a velocity $v = \frac{1}{2}\sqrt{(Lg/2\pi)}$ giving a Doppler shift which is dependent on the square root of ΔF

$$f_s = \sqrt{\frac{g\Delta F}{\pi c}} \tag{7.16}$$

This makes it possible to separate the signature of the internal wave from that of the surface waves as illustrated in Fig. 7.10 and further emphasized in Fig. 7.11.

In Section 6.2 we studied the multifrequency temporal properties of non-dispersive ships relative to that of dispersive ocean gravity waves.

The case of internal waves against a background of surface gravity waves is analogous to the ship/sea surface situation as illustrated in Fig. 7.12.

Here we have plotted the ratio of surface wave Doppler shift and internal wave Doppler shift as a function of radar beat frequency ΔF.

Note that there is only one value of ΔF for which the Doppler ratio is unity, namely $\Delta F = 400$ MHz when $d = 50$ m and $\delta\rho/\rho = 10^{-3}$.

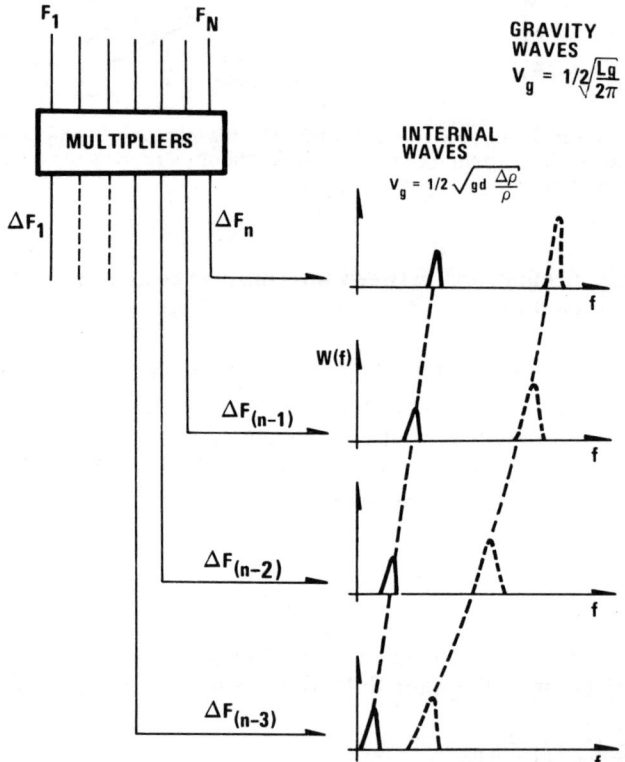

Fig. 7.10 Surface gravity waves are dispersive; long internal waves on a shallow thermocline are not. As a result of this, our adaptive radar allows us to distinguish between the two. There is only one value of ΔF where the Doppler shift caused by the surface wave coincides with that caused by the internal wave

Detection of ship wakes by matched illumination

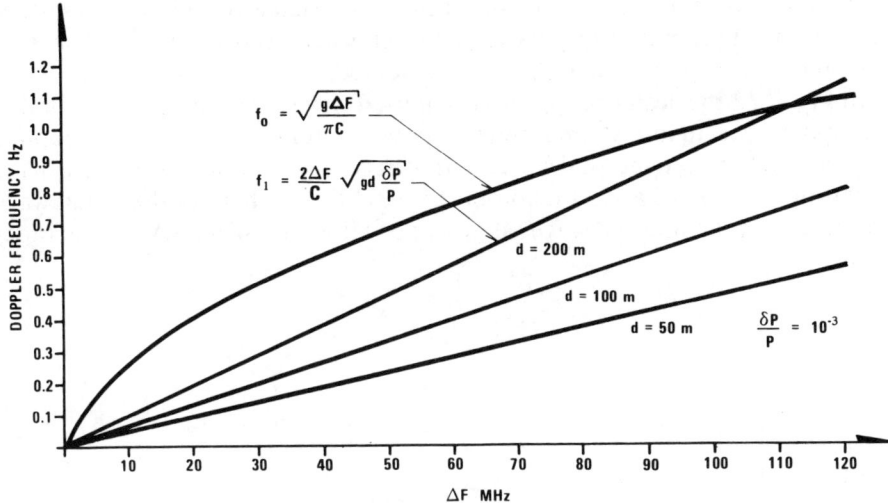

Fig. 7.11 Surface waves are dispersive; internal waves are not. Hence, except for zero ΔF, there is only one value of ΔF that gives interference between the two wave phenomena

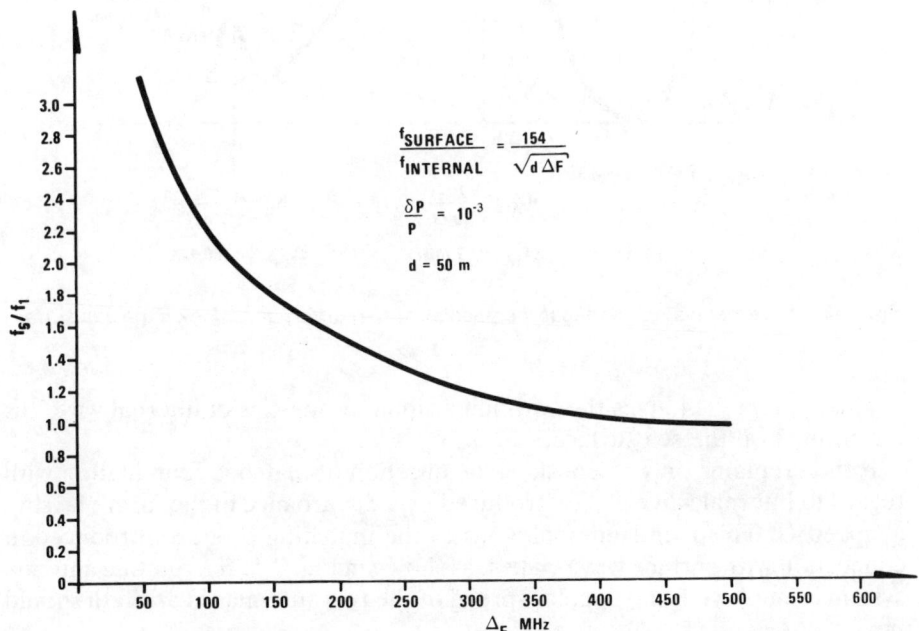

Fig. 7.12 Ratio of Doppler shift caused by the surface wave to that of the internal wave f_s/f_i as a function of ΔF

In summing up this section on footprints of internal waves on the sea surface against a background of dispersive coherent gravity waves and non-coherent irregularities, Figs. 7.13 and 7.14 are presented.

In Fig. 7.13 the scalar properties of internal waves are presented. Note that internal waves manifest themselves at two different scales. The coupling mechanism between internal waves and surface waves shows up at scales in the order of metres (ΔF in the order of 100 MHz) whereas the long wave periodicity of the internal waves shows up at kilometre scales (ΔF in the order of 100 kHz).

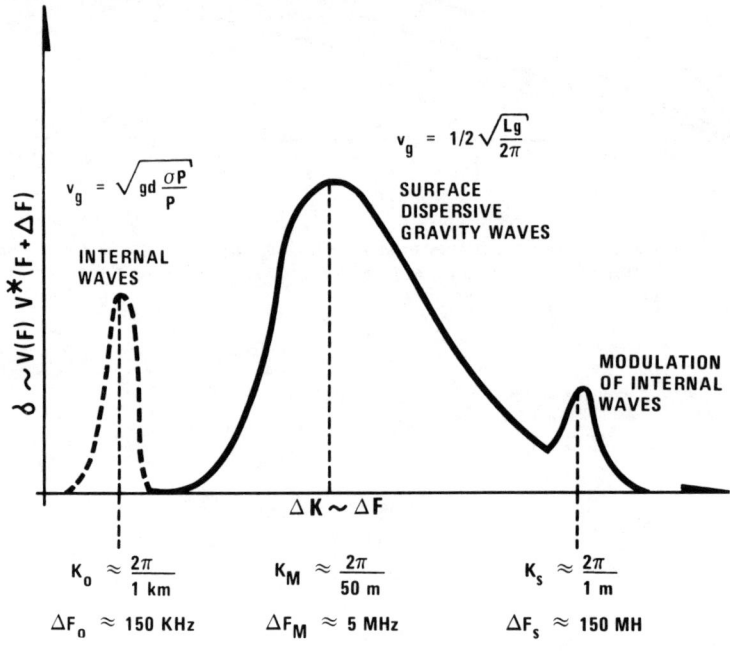

Fig. 7.13 *Internal waves manifest themselves at two different scales on the sea surface*

Finally, Fig. 7.14 gives the two-dimensional properties of internal waves as footprinted on the sea surface.

It then remains only to consider the question of matched illumination with regard to internal wave wakes produced by a disturbance in the form of a ship at speed U. This is indeed analogous to the matching process introduced in connection with surface wave wakes, as shown in Fig. 7.5. If a rotating antenna system is employed, the angular spread of the two antennas in azimuth should be

$$2\psi = 2\cos^{-1}\left\{\frac{\sqrt{[(\Delta\rho/\rho)\,gd]}}{U}\right\}$$

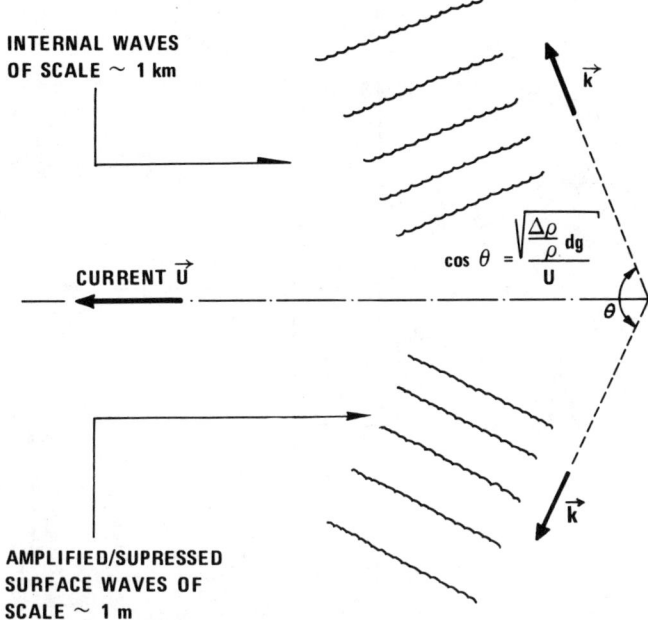

Fig. 7.14 *Two-dimensional footprint of internal waves caused by a ship disturbing the thermocline. Note that the internal wave shows up at two different scales on the surface*

whereas the beat frequency ΔF should be

$$\Delta F = \frac{c}{2L} = \frac{c}{16\pi(\delta\rho/\rho)\,d}$$

By forming the cross-spectrum (mutual coherence) between the signal received at the two antennas as described in Sections 2.5 and 6.3, the influence of the internal wave pattern is enhanced substantially. This is illustrated in Fig. 7.15.

This treatment of the matched illumination concept is based on rather straightforward principles from basic physics. With some background in information theory and in the field of communications the simple matched illumination concept for dispersive and non-dispersive phenomena is rather trivial since it has such clear counterparts in modulation theory.

So far we have in essence considered simple amplitude modulation principles producing two sidebands and carrier frequency. If we are dealing with ocean wave phenomena which cannot be regarded as a set of superimposed periodic wave trains to which amplitude modulated fixed frequency radio waves can be matched, but rather a sea-surface structure which resembles frequency modulated sources, a somewhat more complicated matching scheme will have to be implemented involving frequency modulated (chirp) systems. Although such discussions conceptually are somewhat more

158   Detection of ship wakes by matched illumination

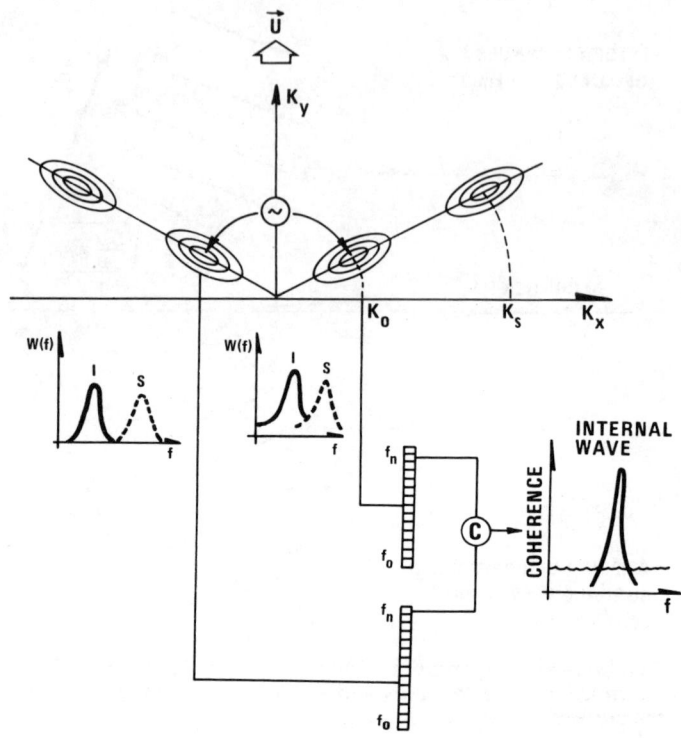

Fig. 7.15   Matched illumination in relation to internal waves caused by a ship at speed U disturbing the thermocline at depth d with a density gradient $\Delta\rho/\rho$

complicated, since the rate of change of illuminating radio frequency shall have to be synchronized through a self-synchronizing (phase-locked loop) system to the dispersive ocean wave train, implementation schemes are well within present state technology involving chirp systems based on surface acoustic wave (SAW) technology. For details regarding such spread spectrum techniques the reader is referred to a recent book by Jens Hjelmstad, from the author's organization, and Reidar Skaug (1985).

References

ALPERS, W. R. and HASSELMANN, K. (1978): 'The two-frequency microwave technique for measuring ocean wave spectra from an airplane or satellite', *Bound. Layer Meteorol.*, **13**, pp. 214–230

ALPERS, W. R. and RUFENACH, C. L. (1979): 'The effect of orbital motions on synthetic aperture radar imagery of ocean waves', *IEEE Trans.*, **AP-27**, pp. 685–690

APEL, J. R., BYRNE, H. M., PRONI, J. R. and CHARNELL, R. L. (1975): 'Observations of oceanic internal and surface waves from the Earth Resources Technology Satellite', *J. Geophys. Res.*, **80**, 6, pp. 865–881

APEL, J. R. and GONZALES, F. I. (1983): 'Nonlinear features of internal waves off Baja California as observed from the SEASAT Imaginary Radar', *J. Geophys. Res.*, **88**, 7, pp. 4459–4466

APPLEBAUM, S. (1976): 'Adaptive arrays', *IEEE Trans.*, **AP-24**, (5), pp. 585–589

BARRICK, D. E. (1972): 'First order theory and analysis of MF/HF/VHF scatter from the sea', *IEEE Trans.*, **AP-20**, pp. 2–10

BASS, F. B. and FUKS, I. M. (1979): 'Wave scattering from statistically rough surfaces' (Pergamon Press)

BATCHELOR, G. K. (1955): 'The scattering of radio waves in the atmosphere by turbulent fluctuations in refractive index', Research Report No. EE262, School of Engineering. Cornell University

BEAL, R. C., DELEONIBUS, P. S. and KATS, I. (eds.) (1981): 'Spaceborne synthetic aperture radar for oceanography' (Johns Hopkins University Press, Baltimore, MD)

BOERNER, W.-M. (1980): 'Use of polarization in electromagnetic inverse scattering', Proc. Int. URSI Symp., München, August 16–29

BRETHERTON, E. P. (1969): 'Waves and turbulence in stratified fluids', *Radio Sci.*, **4**, pp. 1279–1287

BROWN, W. M. and PORCELLO, L. (1969): 'An introduction to synthetic aperture radar', *IEEE Spectrum*, **6**, pp. 52–66

CLIFFORD, S. F. and BARRICK, D. E. (1978): 'Remote sensing of sea state by analysis of backscattered microwave phose fluctuations', *IEEE Trans.*, **AP-26**, pp. 699–705

CRAWFORD, A. B., HOGG, D. C. and KUMMER, W. H. (1959): 'Studies in tropospheric propagation beyond the horizon', *Bell Syst. Tech. J.*, **38**, pp. 1067–1178

DeLOOR, G. P. *et al.* (1974): 'Radar cross sections of vegetation canopies determined by monostatic and bistatic scatterometry', Proc. 9th Symposium Remote Sensing Environment 1457–1466 Ann Arbor, Michigan: University of Michigan, April

DITTEL, R., GJESSING, D. T. and HJELMSTAD, J. (1985): 'Analysis of experimental data from measurements on ships and aircarft against a sea clutter background with special emphasis on the mutual coherency effects', NTNF report TN-70/85

References

DYSTHE, K. B. (1980): 'Havbølger og fysikk', *Fra: Fysikkens Verden*, No. 4, pp. 84–87 and No. 1, 1981, pp. 20–24

ECKERT, J. A., *et al.* (1974): 'Downlooking airborne lidar studies', EPA report, EPA, Las Vegas, Nevada

EOM, H. J. and FUNG, A. K. (1982): 'Scattering coefficients of Kirchhoff surfaces with Gaussian and non-Gaussian surface statistics', Tech. Rep. RSL TR 4601-2, Remote Sensing Laboratory, University of Kansas, Lawrence, Kansas

GABRIEL, W. F. (1976): 'Adaptive arrays – an introduction', *IEEE Proc.*, **64**, p. 2

GJESSING, D. T. (1962): 'On the scattering of electromagnetic waves by non-isotropic inhomogeneities in the atmosphere', *J. Geophys. Res.*, **67**, 3, pp. 1017–1026

GJESSING, D. T. (1964): 'Determination of isotropy properties of the tropospheric permitivity and wind velocity fields by radio-propagation methods', *J. Geophys. Res.*, **69**, 4, pp. 569–582

GJESSING, D. T. (1973): 'On the use of forward scatter techniques in the study of turbulent stratified layers in the troposphere', *Bound. Layer Meteorol.*, **4**, pp. 377–396

GJESSING, D. T. (1978a): 'Remote surveillance by electromagnetic waves for air–water–land' (Ann Arbor Science Publishers)

GJESSING, D. T. (1978b): 'A generalized method for environmental surveillance by remote probing?, *J. Radio Sci.* (March/April)

GJESSING, D. T. (1981a): 'Adaptive techniques for radar detection and identification of objects in an ocean environment', *IEEE J. Ocean Engineer.*, **OE-6**, 1

GJESSING, D. T. (1981b): 'Adaptive radar in remote sensing' (Ann Arbor Science Publishers)

GJESSING, D. T. and BORRESEN, J. (1968): 'The influence of an irregular refractive index structure on the spatial field-strength correlation of a scattered radio wave', IEE Conf. Proc. 48, September

GJESSING, D. T., HAMRAN, S.-E., HJELMSTAD, J. and AARHOLT, E. (1985): 'Identification of species of vegetation and measurement of stress factors by the signature domain texture, motion pattern and colour', NTNF rep. TN-65/85 (in Norwegian)

GJESSING, D. T. and HJELMSTAD, J. (1986): 'Ocean waves and turbulence as observed with an adaptive coherent multifrequency radar', *J. Geophysical Research Oceans*, January 1986

GJESSING, D. T., HJELMSTAD, J and LUND, T. (1982): 'A multifrequency adaptive radar for detection and identification of objects. Results of preliminary experiments on aircraft against a sea-clutter background', *IEEE Trans.*, **AP-30**, 3, pp. 351–365

GJESSING, D. T., HJELMSTAD, J. and LUND, T. (1983): 'Optimum detection techniques in relation to shape and size of objects, motion pattern and material composition, *in* BOERNER, W.-M. *et al.*, (ed.) 'Inverse methods in electromagnetic imaging, Part 2' (D. Reidel Publishing Company), pp. 871–905

GJESSING, D. T., HJELMSTAD, J. and LUND, T. (1985): 'Directional ocean wave spectra as observed with multifrequency continuous wave radar', Review paper in *International Journal of Remote Sensing*, **7**, pp. 979–1008

GJESSING, D. T. and IRGENS, F. (1964a): 'On scattering of electromagnetic waves by a moving trophospheric layer having sinusoidal boundaries', *IEEE Trans.*, **AP-12**, 1, pp. 51–62

GJESSING, D. T. and IRGENS, F. (1964b): 'Scattering of radio waves by moving atmospheric rippled layers: a simple model experiment', *IEEE Trans.*, **AP-12**, 6, pp. 703–709

GJESSING, D. T., JESKE, H. and KLINT-HANSEN, N. (1969): 'An investigation of the tropospheric fine scale properties using radio, radar and direct methods', *J. Atmosph. Terr. Phys.*, **31**, pp. 1157–1182

GJESSING, D. T., KJELAAS, A. G. and NORDO, J. (1969): 'Spectral measurements and atmospheric stability', *J. Atmosph. Sci.*, **26**, 3, pp. 462–468

GJEVIK, B. (1983): 'Internal ship-wave pattern in a two-layered sea', Internal Report, Dept. of Mechanics, Oslo University, May

GJEVIK, B. (1984): 'SLAR-detection of long-crested internal waves in Skagerak', *Naturen*, **6**, pp. 209–215

GJEVIK, B. and MARTINSEN, T. (1978): 'Three-dimensional lee-wave pattern', *Quart. J. R. Mat. Soc.*, **104**, p. 947–957
GRØNLIE, Ø., BRODTKORB, D. C. and WØIEN, J. (1984): 'MIROS – A microwave remote sensor for the ocean surface', *Norwegian Maritime Research*, **12**, No. 3, pp. 24–28
HAGFORS, T. (1959): 'Investigation of the scattering of radio waves at metric wavelengths in the lower ionosphere', *Geophysica Norwegica*, **21**, 2
HAGFORS, T. (1961): 'Some properties of radio waves reflected from the moon and their relation to the lunar surface', *J. Geophys. Res.*, **66**, 3, pp. 777–785
HASSELMANN, K. et al (1973): 'Measurements of wind–wave growth and swell decay during the Joint North Sea Wave Project (JONSWAP)', *Ergänzungsheft zur Deutschen Hydrographischen Zeitshrift*, Reihe **A/80**, 12
HJELMSTAD, J. (1983): 'On the use of communication concepts and technique for remote sensing', Proc. URSI Commission F 1983 Symposium, Louvain, Belgium, June (ESA SP-194), pp. 381–390
HJELMSTAD, J. and SKAUG, R. (1985): 'Spread spectrum communications with military applications' (Peter Perigrinus)
ISHIMARU, A. (1978): 'Wave propagation and scattering in random media' (Academic Press, New York)
JACKSON, F. C. (1980): 'Directional spectra of ocean waves from microwave backscatter', General Electric Co., Space Division Report
JACOBSEN, S. K. (1985): 'A study of scattering mechanisms from a rough sea surface using multifrequency radar', M.Sc. Thesis, University of Tromsø
JONES, L. and WEISSMAN, D. E. (1981): 'The two frequency microwave scatterometer measurements of ocean wave spectra from an aircraft', Oceanography from Space. Proc. COSPAR/SCOR/IUCRM Symposium, GOWER, J. F. R. (ed.) Plenum Press, New York
KJELAAS, A. G., NORDAL, P. E. and BJERKESTRAND, A. (1977): "Multiwavelength scintillation effects in long-path CO_2 laser absorbtion spectrometer" Proc. URSI, Com. F. Symposium, La Baule, France 1977
KLEMAN, J. (1985): 'The spectral reflectance of coniferous tree stands and of barley influenced by stress. An analysis of field measured spectral data', Report No. A173, Naturgeografiska Institutionen vid Stockholms Universitet
KWOH, D. S. W. and LAKE, B. M. (1983): 'Microwave backscattering from short gravity waves. A deterministic coherent and dual-polarized laboratory study', Office of Naval Research Coastal Sciences Program, California
LUMLEY, J. L. (1967): 'Theoretical aspects of research on turbulence in stratified flows: *In* 'Atmospheric turbulence and radio wave propagation' YAGLOM, A. M. and TATARSKY, V. I. (eds.) (Moscow, Publishing House NAUKA), pp. 105–112
MEGAW, E. C. S. (1957): 'Fundamental radio scatter propagation theory', *Proc. IEEE*, Monog. No. 236R
PANOFSKY, H. A. and McCORMICK, R. A. (1960): 'The spectrum of vertical velocity near the surface', *Quart. J. Roy Met. Soc.*
PHILLIPS, O. M. (1957): 'On the generation of waves by turbulent wind, *J. Fluid Mech.*, **2**, 5, p. 417
PHILLIPS, O. M. (1969): 'The dynamics of the upper ocean' (Cambridge University Press) p. 159
PIERSON-MOSKOWITZ (1964): 'A proposed spectral form for fully developed wind seas based on the similarity of S.A. Kitaigorodski', *J. Geophys. Res.*, **69**, pp. 5181–5190
PLANT, W. J. (1977): 'Studies of backscattered sea return with a CW dual-frequency X-band radar', *IEEE Trans.*, **AP-25**, 1
PLANT, W. J. and SCHULER, D. L. (1980): 'Remote sensing of the sea surface using one and two frequency microwave', *Radio Sci*, **15**, 3, pp. 605–615
RAMAMONJIARISOA, A. (1974) "Contribution a l'étude de la structure statistique et des mécanismes de génération des vagues de vent. Theses, Université de Provence No. A.O. 10, 023

RAMAMONJIARISOA, A., et al. (1978): 'Laboratory studies in wind–wave generation, amplification and evolution' In 'Turbulent fluxes through the sea-surface: wave dynamics and prediction' FAURE, A. and HASSELMAN, K. (eds.) (New York, Plenum Publishing Corporation)

RATCLIFFE, J. A. (1965): 'Some aspects of diffraction theory and their application to the ionosphere', Phys. Soc. Rep. Prog. Phys., **19**, p. 188

SCHULER, D. L. (1978): 'Remote sensing of directional wave spectra and surface currents using a microwave dual-frequency radar', Radio Sci., **13**, pp. 321–33, March–April

SHUCHMAN, R. A., KASISCHKE, E. S., LYZENGA, D. R. and KLOOSTER, A., Jun. (1983): 'SAR ship wake signatures', Ann Arbor, MI 48107, Radar Division, Environmental Research Institute of Michigan

STRATTON, J. A. (1941): 'Electromagnetic theory' (McGraw-Hill, New York)

TATARKSY, V. E. (1961): 'Wave propagation in a turbulent medium' (Translated by R. A. Silverman, McGraw-Hill, New York)

ULABY, F. T., BATLIVALA, P. P. and DOBSON, M. C. (1978): 'Microwave backscatter dependence on surface roughness, soil moisture, and soil texture: Part 1 – Bare soil', IEEE Trans., **GE-16**, (10), p. 286–295

VALENZUELA, G. R. (1978a): 'Theories for the interaction of electromagnetic and oceanic waves – A review', Bound. Layer Meterorol., **13**, pp. 61–85

VALENZUELA, G. R. (1978b): 'Scattering of electromagnetic waves from the ocean. Surveillance of environmental pollution and resources by electromagnetic waves', Proc. NATO Advanced Study Institute, Spåtind, Norway, 9–19 April 1978, LUND, T. (ed.)

VAN TREES, H. L. (1971): 'Detection estimation and modulation theory, Part III/1' (New York, John Wiley)

WATERMAN, A. T., GJESSING, D. T. and LISTON, C. L. (1961): 'Statistical analysis of Transmission data from a simultaneous frequency- and angle scan experiment', Contribution to the URSI Spring Meeting, Washington DC

WEISSMAN, D. E. and JOHNSON, J. W. (1977): 'Dual frequency correlation radar measurements of the height statistics of ocean waves', IEEE J. Ocean Engineer., **OE-2**, pp. 74–83

WEISSMAN, D. E., JOHNSON, J. W. and RAMSEY, J. T. (1982): 'The delta-K ocean wae spectrometer: Aircraft measurements and theoretical system analysis', Proc. IGARSS –82

WELSH, E. J. KENNEDY, J. E. and ULIANA, E. A. (1979): 'The surface contour radar. A unique remote sensing instrument', IEEE Trans., **MTT-27**, 12

WIENER, N. (1930): 'Generalized harmonic analysis', Acta Math., **55**

WOODS, J. D. (1969): 'On Richardson's number as a criterion for laminar-turbulent-laminar transition in the ocean and atmosphere', Radio Sci, **4**, 12, pp. 1289–1298

WRIGHT, J. W. (1968): 'A new model for sea clutter', IEEE Trans., **AP-16**, pp. 217–223

WRIGHT, J. W. PLANT, W. J., KELLER, W. C. and JONES, L. (1980): 'Ocean wave–radar modulation transfer functions from the West Coast Experiment', J. Geophys. Res., **85**, C9, pp. 4657–4966

Bibliography

ALPERS, W. and HASSELMAN, W. (1978): 'The two-frequency microwave technique for measuring ocean-wave spectra from an airplane or satellite', *Bound. Layer Meteor.*, **13**, pp. 215–230

ALPERS, W. E., ROSS, D. B. and RUFENACH. C. L. (1981): 'On the detectability of ocean surfaces by real and synthetic aperture radar', *J. Geophys. Res.*, **86**, pp. 6481–6498

ALPERS, W. R. and RUFENACH, C. L. (1979): 'The effect of orbit motions on synthetic aperture radar imagery of ocean waves', *IEEE Trans.*, **AP-27**, pp. 685–690

AUSHERMAN, D. A., HALL, W. D., LATTA, J. N. and ZELENKA, J. S. (1975): 'Radar data processing and exploitation facility', Proc. IEEE International Radar Conference, Washington DC

BECKAMAN, P. and SPIZZICHINO, A. (1963): 'The scattering of electromagnetic waves from rough surfaces' (New York, Macmillan)

BENIGNUS, V. A (1969): 'Estimations of coherency spectrum and its confidence interval using fast Fourier transform', *IEEE Trans.*, **AU-17**, (2)

BENNETT, C. L. (1980): 'Inverse scattering: time-domain in solutions via integral equations' In 'NATO advanced study institute on theoretical methods for determining the interaction of electromagnetic waves with structures' SKWIREYNSKI, J. K. (ed.) (Sijphoff and Noordhoff, Amsterdam)

BENNETT, H. F. (1963): 'Specular reflectance of aluminized ground glass and the height distribution of surface irrgularities', *J. Opt. Soc. Amer.*, **53**, pp. 1389–3194

BOERNER, W.-M. (1978): 'State of the art review on polarization in electromagnetic inverse scattering', Rep. 78-3, Communications Laboratory, University of Illinois, Chicago

BOERNER, W.-M. (1980): 'Polarization microwave holography: an extension of scalar to vector holography', 1980 Int. Optics Computing Confl, SPIE's Tech. Symp. East, Washington, DC, Session 3B, paper 231–33, pp. 188–198

BOERNER, W.-M. (1982): 'Polarization control in radar meteorology', URSI Commission F/IEE. Open symposium, Bournemouth, U.K.

BOERNER, W.-M., EL-ARINI, M. B., CHAN, C. Y., SAATCHI, S. S., IP, W. S., MASTORIS, P. M. and FOO, B. Y. (1981): 'Polarization utilization in radar target reconstruction', Univ. Illinois at Chicago Circle, Tech. Rep. C1-EMID-NANRAR-81-01

BOJARSKI, N. N. (1967): 'Three dimensional electromagnetic short pulse inverse scattering', Special Projects Lab., Rep. AS845, Syracuse Univ. Res. Corp, Syracuse, N.Y.

BROWN, G. S. (1978): 'Backscattering from a Gaussian-distributed perfectly conducting rough surface', *IEEE*, **AP-26**, 3, p. 472

COX, D. C. and WATERMANN, A. T., Jun. (1968): 'Phase and amplitude measurements of transhorizon microwaves with a multidata-gathering antenna array', AGARD Conference Proceedings 37, pp. 18–1 to 18–6

CROMBIE, D. D. (1971): 'Backscatter of HP radio waves from the sea in WAIT, J. R. (ed.) 'Electromagnetic probing in geophysics' (Boulder, Colorado, The Golden Press), pp. 131–162

DESCHAMPS, G. A. (1951): 'Geometrical representation of the polarization of a plane electromagnetic wave', *Proc. IRE*, **39**, pp. 543–538 (see also 'A hyperbolic protractor for microwave impedance measurements and other purposes', Federal Telecommunication Labs., ITT, Nutley, NJ., 1953)

DOBSON, M. C. and ULABY, F. T. (1981): 'Microwave backscatter dependence on surface roughness, soild moisture, and soil texture: Part III – Soil terrain', *IEEE Trans.*, **GE-19** (1), pp. 51–61

EKLUND, F. and WICHERTS, S. (1968): 'Wavelength dependence of microwave propagation far beyond the radio horizon', *Rdio Sci.*, **3** (11), pp. 1068–1974

ENOCHSEN, L. and GOODMAN, N. R. (1965): 'Gaussian approximations to the distribution of sample coherence', AFFDL TR-65-57, Research and Technology Division, ASCF, Wright-Patterson AFB, Ohio (Feb.)

FEHLABER, L. (1966): 'Diversity Abstand auf Scatter Strecken im Frequenzbereich zwischen 1 GHz and 10 GHz', *Tech. Vericht.*, 5581

FEHLABER, L and GROSSKOPF, J. (1968): 'Messung der Gweinnminderung bei 1715 MHz aur einer 409 km langen Scatter-Versuchsstrecke', *Fernmeldetechnisches Zentralamt*, FTZ-A455

FUNG, A. K. (1967): 'Theory of cross-polarized power returned from a random surface', *Appl. Sci. Res.*, **18**, pp. 50–60

FUNG, A. K. and CHEN, M. F. (1983): 'Scattering from a perfectly conducting random surface extinction method', Tech. Rep. RSL TR 592-3, Remote Sensing Laboratory, University of Kansas, Lawrence, Kansas

FUNG, A. K. and LEE, K. K. (1983): 'Variation of sea wave spectrum with wind speed', Digest of IGARSS '83, San Francisco, California, August 31–September 2

GJESSING, D. T. (1964): 'An experimental determination of the spectrum of permittivity and air velocity fluctuations along a vertical direction in the troposphere using radio propagation methods', *J. Atmosph. Terr. Phys.*, **26**, 2

GJESSING, D. T. (1967): 'Radiophysical aspects of irregular structure in the atmosphere' in 'Atmospheric turbulence and radio wave propagation' YAGLOM, A. M. and TATARKSY, V. I. (eds.) (Moscow, USSR, Publishing House Nauka)

GJESSING, D. T. (1968): 'Scattering of radio waves from regular and irregular time varying refractive-index structures in the troposphere', AGARD Conference Proceedings, vol. 37, pp. 15–1 to 15–17

GJESSING, D. T. (1969): 'Atmospheric structure deduced from the forward scatter wave propagation experiments', *Radio Sci.*, **4**, (12), pp. 1195–1210

GJESSING, D. T. (1978): 'Target detection and identification methods based on radio- and optical waves', AGARD Lecture Series No. 93

GJESSING, D. T. (1979): 'Environmental remote sensing. Part I: Methods based on scattering and diffraction of radio waves', *Phys. Technol.*, **10**

GONZALEZ, F. I., BEAL, R. C., BROWN, W. E., Jun., DeLEONIBUS, D. S., GOWER, J. F. R., LICHY, D., ROSS, D. B., RUFENACH, C. L., SHERMAN, J. W., III, and SHUCHMAN, R. A. (1979): 'Seasat synthetic aperture radar: ocean wave detection capabilities', *Science*, **204**, pp. 1418–1421

GROSSKOPF, J. (1968): 'Investigation of the receiving field for scatter propagation', AGARD Conf. Proc. 37, pp. 22–1 to 22–11

HASSELMANN, K. (1966): 'Feynman diagrams and interaction rules for wave–wave scattering', *Rev. Geophys.*, **4**, pp. 1–32

HUYNEN, J. R. (1970): 'Phenomenological theory of radar targets', Ph.D. thesis, tech. Univ. Delft, Druckerij Bronder-Offset N. V. Rotterdam, Netherlands

JANES, H. B. and THOMPSON, M. C. (1973): 'Comparison of observed and predicted phase-front distortion in line-of-sight microwave signals', *Trans. IEE*, **AP-21**, (2), pp. 263–311
JEFFREYS, H. (1925): 'On the formations of water waves by wind', *Proc. Roy. Soc. A. London*, **107**, p. 189
JENNISON, R. C. (1961): 'Fourier transforms and convolutions for the experimentalist' (Pergamon Press)
JESKE, H. G. (ed.) (1976): 'Atmospheric effects on radar target identification and imaging', Proc. NATO ASI. Goslar. Dordrecht, Holland: D. Reidel
KATZIN, M. (1957): 'On the mechanisms of radar sea clutter', *Proc. IRE*, **45**, pp. 44–54 (Jan)
KENNAUGH, E. M. (1949–1954): 'Effects of type of polarization on echo characteristics', Ohio State University, Antenna Laboratory, Columbus, OH, Reports 389-1 to 389–24
KENNAUGH, E. M. (1950): 'Effects of type of polarization on echo characteristics monostatic case', Ohio State University Antenna Laboratory, Columbus, OH 43212, Report 389-1, Sep. 16, 1949 and Report 389-4, June 16
KJELAAS, A. G., NORDAL, P. E. and BJERKESTRAND, A. (1977): 'Multi-wavelength scintillation effects in a long-path CO_2 laser absorption spectrometer', Proc. URSI Comm F Symposium, La Baule, France, April 28–May 6
LEE, R. W. (1974): 'Remote probing using spatially filtered apertures', *J. Opt. Sec. Amer.*, **64**, p. 1295
LEE, R. W. and HARP, J. C. (1969a): 'Wek scattering in random media, with applications to remote probing', *Proc. IEEE*, **57**, pp. 375–406
LEE, R. W. and HARP, J. C. (1969b): 'Weak scattering in random media', *J. IEEE*, **57**, p. 375
LONG, M. W. (1975): 'Reflectivity of land and sea' (Lexington, MA, Lexington Books)
LONGUET-HIGGINS, M. S. (1963): 'The generation of capillary waves by steep gravity waves', *J. Fluid Mech.*, **16**, pp. 138–139
NATHANSON, R. D. (1969): 'Radar design principles' (New York, McGraw-Hill)
OPENHEIM, A. V. and SCHAFER, R. W. (1975): 'Digital signal processing' (New Jersey, Prentice-Hall)
OTNES, R. K. and ENOCHSEN, L. (1972): 'Digital time series analysis' (John Wiley)
OTNES, R. K. and ENOCHSEN, L. (1978) 'Applied time series analysis. vol. 1: Basic techniques' (John Wiley)
PANOFSKY, W. K. H. and PHILLIPS, M. (1955): 'Classical electricity and magnetism' (Addison-Wesley)
PAPOULIS, A. (1962): 'The Fourier integral and its applications' (McGraw-Hill)
PAPOULIS, A. (1979): 'Signal analysis' (McGraw-Hill)
PARZEN, E. (1961): 'Mathematical considerations in the estimation of spectra', *Technometrics*, **3**, pp. 167–190
PHILLIPS, O. M. (1958): 'The equilibrium range in the spectrum of wind generated waves', *J. Fluid Mech.*, **4**, 4, p. 426
PHILLIPS, O. M. and BANNER, M. L. (1974): 'Wave breaking in the presence of wind drift and swell', *J. Fluid Mech.*, **66**, pp. 625–640
POELMAN, A. J. (1979): 'Reconsideration of the target detection criterion based on adaptive antenna polarization', *Tijdschr. Nederlands Electron. Radiogenootschap*, **44**, pp. 93–106
POND, S. and PICKARD, L. (1976): 'Introductory dynamic oceanography' (Pergamon Press)
RABINER, L. R. and GOLD, B. (1975): 'Theory and application of digital signal processing' (Prentice-Hall, New Jersey)
RAMAMONJIARISOA, A., *et al.* (1978): 'Observations de la vitesse de propagation des vagues engindrees par let vent au large', *C.R. Acad. Sci., Paris*, September 18, p. 287
RANEY, R. K. (1971): 'Synthetic aperture imaging radar and moving targets, *IEEE Trans.*, **AES-7**, pp. 499–505
RANEY, R. K. (1980): 'SAR response to partially coherent phenomena', *IEEE Trans.*, **AP-28**, pp. 777–787

RAWSON, R., SMITH, F. and LARSON, B. (1975): 'The ERIM X- and L-band dual polarized radar', IEEE 1975 International Radar Conference, New York, p. 505

RICE, S. O. (1944): 'Mathematical analysis of random noise', *Bell Syst. Tech. J.*, **23**, p. 282

RICE, S. O. (1951): 'Reflection of electromagnetic waves from slightly rough surfaces', *Comm. Pure Appl. Math.*, **4** (2/3), pp. 351–378

SANCER, M. I. (1969): 'Shadow-corrected electromagnetic scattering from a randomly rough surface', *IEEE Trans.*, **AP-17**, pp. 577–597

SCHNELL, A. C. (1979): 'Programs for digital signal processing' (John Wiley)

SCHULE, J. J., SIMPSON, L. S. and DELEONIBUS, P. S. (1971): 'A study of fetch limited wave spectra with an airborne laser', *J. Geophys. Res.*, **76**, pp. 4160–4171

SINCLAIR, G. (1948): 'Modification of the radar range equation for arbitrary targets and arbitrary polarization', Antenna Lab., Electrosci. Lab. Report 302,19, Ohio State University, Columbus, OH

SKOLNIK, M. I. (1962): 'Introduction to radar systems' (McGraw-Hill, New York) (see also SKOLNIK, M. I. (1978): Radar handbook' (McGraw-Hill, New York)

STEARNS, S. D. (1975): 'Digital signal analysis' (Hayden Book Co. Inc., New Jersey)

STIEFFEL, E. (1959): 'Numerical methods of Tchebycheff approximation on numerical approximation' (R. E. Langer (eds), University of Wisconsin Press), pp. 217–232

STRATTON, J. A. (1941): 'Electromagnetic theory' (McGraw-Hill)

TOMIYASU, K. (1971): 'Short pulse wide-band scatterometer ocean surface signature', *IEEE Trans.*, **GE-9**, pp. 175–177

ULABY, F. T., MOORE, R. K. and FUNG, A. K. (1982): 'Microwave remote sensing, vol. 2' (Addison-Wesley), Chapter 12

VALENZUELA, G. R., et al. (1977): 'Modulation of short gravity-capillary waves by longer scale periodic flows', Technical Note, Naval Research Laboratory

VALENZUELA, G. R., LAING, M. B. and DALEY, J. C. (1971): 'Ocean spectra for the high-frequency waves as determined from airborne radar measurements', *J. Marine Res.*, **29**, pp. 69–84

VANDEN-BROCK, J. (1974): 'Mécanique es vagues de grande amplitude', Thesis for Ingen. Phys., Université de Liege

WAIT, J. R. (1964): 'A note on VHF reflection from a tropospheric layer', *Radio Sci.*, **7**, pp. 847–878

WEISSMAN, D. E. (1973): 'Two frequency radar interferometry applied to the measurement of ocean wave height', *IEEE Trans.*, **AP-21**, 5 (Sep)

WEISSMAN, D. E. and JOHNSON, J. W. (1979): 'Rough surface wavelength measurement through self mixing of Doppler microwave backscatter', *IEE Trans.*, **AP-27**, pp. 730–737

WELCH, P. D. (1970): 'The use of fast Fourier transform for the estimation of power spectra', *IEEE Trans*, **AU-15**

WHALEN, A. D. (1971): 'Detection of signals in noise' (Academic Press, New York)

WHITHAM, G. B. (1974): 'Linear and nonlinear waves' (John Wiley)

WRIGHT, J. W. (1966): 'Backscattering from capillary waves with application to sea clutter', *IEEE Trans.*, **AP-14**, 6, pp. 749–754

WU, S. T. and FUNG, A. K. (1972): 'A noncoherent model for microwave emissions and backscattering from the sea surface', *J. Geophys. Res.*, **77**, (30), pp. 5917–5929

YUEN, C. K. and FRASER, D. (1979): 'Digital spectra analysis' (Pitman Advanced Publishing Program, London)

Index

absorbing surfaces, 55
absorption spectra, 54
absorption, 55
acid precipitation, 84
adaptive polarization, 71
additive noise, 60, 3
adequate description, 37
adiabatic invariance, 96
air pollution, 57, 84
air/sea phenomena, 92
altimeter, 115
amplitude distribution, 48
amplitude modulating, 20, 34, 60, 77
amplitude scintillations, 46
analogue processing, 86
angle of arrival spectrum, 70
angular distribution, 25, 110, 127
angular power spectrum, 27, 34, 70, 102
angular power distribution, 27
angular resolution, 103, 114
angular spread, 104, 106, 107
anisotropic irregularity structure, 108
anisotropy, 110
antenna aperture, 102, 110
antenna array, 28
antenna beamwidth, 104
antenna theory, 27, 102
apriori information, 74
array processors, 74
association plot, 143
autocorrelation function, 34, 60
autocorrelation, 15
autocovariance, 60
autopilot, 77
axis of symmetry, 127
azimuth resolution, 102

ball and spring system, 56
bandwidth measurements, 93, 111

bandwidth, 4, 5, 19, 20, 80, 93
beat channels, 19
beat frequency matching, 10
biological processes, 89
boundary layer wind field, 92
bow wave envelope, 150
Bragg angle, 120
Bragg scattering, 112
breaking waves, 111, 113
broadband radar, 23
broadside array, 6, 28
bulk scattering cross-section, 100

capillary resonance phenomena, 112
capillary wave coupling, 113
capillary wave structure, 92
capillary waves, 93, 111
charge coupled device, 86
chemical agents, 57, 61
chemical compound, 61
chemical substances, 56
chirp FM/CW, 83, 117
co-spectrum, 53
coding techniques, 39
coherence filtering, 52
coherence function, 51, 61
coherency of the ocean wave, 108
coherent integration, 74
coherent microwaves, 137
coherent radar, 35, 16
coherent wave motion, 92, 95, 125
colour distribution, 84
colour measuring satellites, 84
communication systems, 23
complex amplitude, 47
complex autocorrelation, 22
complex correlation, 16
complex scattering matrix, 39
complex wave-number plane, 148

compressible wave pattern, 139
constructive back-scatter, 101
constructive interference, 9, 17, 102
convolution, 60, 104
corner reflectors, 77, 80
correlated frequencies, 9
correlation time, 46
correlation distance, 15, 23
correlation function, 20
coupling mechanisms, 94
covariance function, 117
crop yield, 89
cross-correlation function, 20
cross-covariance function, 53, 60
cross-polarized returns, 39
cross-power, 53
cross-spectrum, 53, 140, 141
cross-spectrum analysed, 53
cross-spectrum coherence, 51, 52, 89
curvature of the wavefront, 24
cusp wave, 146, 150
cutoff wave length, 99

damping factors, 111
decorrelation time of the target, 48
delay function, 11, 20, 22, 69, 76, 93, 101, 110
delta functions, 80
density gradients, 92, 95
density profile, 145
destructive interference, 9, 77
deterministic description, 85
deterministic molecular absorption spectrum, 50
deterministic signal component, 58
diatomic molecule, 56
diffuse back-scattered component, 8
diffuse non-specular scattering, 101
diffuse scattering, 48, 80
Dirac delta distribution, 101
directional patterns of internal waves, 152
directional resolution, 92
directional spectra, 120, 127
directional wave spectrum, 4
directional properties, 146
dispersion relationship, 16, 76, 114, 117
dispersive ocean gravity waves, 154
dispersive properties, 51
dispersive sea surface, 77
dispersive sea, 139
dispersive surface waves, 146
dispersive waves, 117
dispersiveness of ocean waves, 137
dissipation mechanisms, 98
distribution in depth, 21

dominant target scales, 134
dominating interface, 144
Doppler broadening, 47, 133
Doppler distortions, 3
Doppler effects, 45
Doppler filter, 19, 73, 133
Doppler frequency, 123
Doppler frequency shift, 49
Doppler relationship, 70
Doppler shift, 12, 32, 95, 114, 116, 133
Doppler spectrum, 16, 32
Doppler trajectory, 35
drift velocity, 53, 54
driving wind, 99
dual polarization horn antenna, 73
dynamic image, 89
dynamic properties of the sea, 94

elementary waves, 49
elliptic polarizations, 71
energy levels, 56
environmental surveillance, 4
ergodic, 14
error rate, 4
estimates of rigidity, 140
experimental spectra, 100
exponential shadowing, 21, 23
exponential truncation, 104

fading, 46, 49
fall time, 115
fetch, 106, 107
field-strength distribution, 26
filter function, 104
fingerprints, 57
first-order matching, 64
fluid dynamics, 141
flutter, 75
FM chirp, 115
focusing effects, 92
focusing, 111
footprint of a local wind, 108
footprint, 104, 144
forest signature library, 88
fossilized turbulence, 107
four-dimensional hologram, 28
four-dimensional irregularity spectrum, 28
four-dimensional space-time problem, 55
four-dimensional spectrum, 93
Fourier analysis, 63
Fourier components, 8, 101
Fourier theory, 8
frequency covariance function, 38, 41 119

Index

frequency covariance, 71, 75
frequency difference matching, 110
frequency domain, 6
frequency modulated sources, 157
frequency modulation, 58, 116
frequency selective absorption, 56
Fresnel zone, 11, 47
Froude number, 151
frequency synthesizers, 74
fully developed seas, 98, 99

Gaussian distribution, 16, 75, 104
Gaussian ship, 132
geometry of wave pattern, 144
grating, 47
gravity wave structure, 110, 112
gravity wave, 54, 76, 92, 95, 135
greying of the sea surface, 113
grid of trees, 112
group velocity, 95, 147
gust pattern, 107

height distribution, 85
HF radar, 114
hologram, 29, 88
holographic methods, 19
Hooke's law, 56
human perception, 4
hydrodynamic mechanisms, 93

idealized tree, 85
identification capability, 29
identification potential, 52
imperfect corner reflectors, 131
impulse response, 20, 60
incoherent components, 125
incoherent irregularity structure, 108
incoherent motion, 92, 107
incoherent scattering elements, 50
incoherent signal, 60
incoherent turbulence, 111
incoherent velocity components, 117
information bandwidth, 63
information theory, 157
infra-red waves, 56
integration period, 19
interaction mechanisms, 100
interferents, 62, 63
interferogram, 9, 34
internal waves, 92, 96, 97, 144, 149, 150, 154
intervening propagation medium, 3
inverse Fourier transform, 28
irregularity pattern, 92

irregularity scale, 95
irregularity spectrum, 92, 101, 125
irregularity structure, 47, 95

K matching, 10
K-space signature, 133
Kalman filtering, 7, 69
Kelvin envelope, 146
Kelvin wake, 1

leaf/needle vibration frequency, 91
leaf vibration, 89
linear scanning, 64
longitudinal distribution, 24, 69, 80
longitudinal resolution, 30
lossy surface, 56
lunar surface, 14

macro-scale phenomena, 55
maritime surveillance, 4
matched frequencies, 136
material composition, 55
matrix antenna, 6, 68
maximum entropy, 7, 69
microprocessors, 74
microwave illuminator, 93
microwave transmit/receive units, 73
modified sweep function, 66
modulating wave, 20
modulation theory, 157
modulation transfer function, 93, 112
molecular absorption lines, 56
molecular composition, 56
molecular resonance, 55, 56
molecular structure, 56, 61
molecular surface structure, 56
monostatic imaging, 37
monostatic radar, 36
most significant wave height, 93
motion pattern analysis, 90, 91
motion pattern matching, 134
motion pattern signatures, 137
motion pattern, 69, 70, 86, 89, 95
multi-domain adaptive radar, 38
multifrequency investigations, 112
multifrequency synthetic aperture, 29
multipath fading, 20
multipath mechanisms, 49
multipath phenomena, 56
multiplicative noise, 3, 5, 60
multisensor data fusion, 7, 69
multispectral colour, 84
mutual coherence, 50, 52, 75, 84, 89, 138

170   Index

needle dipoles, 85, 88, 112
noise bandwidth, 19
noise contribution, 63
noise factor, 19
non-co-operative IFF, 138
non-coherent components, 60
non-dispersive process, 54
non-dispersive internal waves, 146
non-dispersive phenomena, 117, 157
non-dispersive ship, 77, 154
non-dispersive wave phenomena, 92
non-periodic delay functions, 132
non-rigid scattering object, 28
normalized correlation, 79
Norway spruce, 85

object classification, 4
object recognition, 36
object rigidity, 71
ocean background, 51
ocean surface structure, 95
ocean surface, 92, 193
ocean wave spectra, 98
ocean waves, 53
optimum parameter estimation, 7
orbital motion, 111
orbital velocity, 116
oriented dihedrals, 42
orthogonal dimensions, 3
orthogonal directions, 30
orthogonal domains, 4
orthogonal excitation, 37
orthogonal response, 37
orthonormal Dirac delta-functions, 37
oscillating modes, 56

parallel processing, 19
pattern signature, 77
penetration depth, 22
perception capability, 4
periodic footprint, 92
phase angle, 54
phase front, 6, 82, 102
phase matching, 108
phase perturbations, 3
phase velocity, 16, 54, 95, 98, 136, 147
phase-locked, 8
phasefront of ocean wave, 102
phasefront of radio wave, 102
pitch, 135
Planck's law of radiation, 1
plane ocean wave, 102
planeness parameter, 105

polarimetric capability, 38
polarimetric methods, 36
polarimetric multifrequency radar, 42
polarimetric radar, 38
polarimetric techniques, 2
polarimetry polarization combinations, 42
polarization measurements, 71
polarization properties, 2, 69
polarization sensitive scattering, 72
polarization signatures, 36
polarization transversal plane, 37
polarization 36, 69, 83
potential forces, 56
power spectrum, 21, 22, 60, 89, 117
precipitation, 113
PRF, 19
pulse compression technique, 115
pulse compression, 83, 117
pulse radar, 19
pulse width, 19
pulsed radar, 110

Q-factor, 16
quadrature spectrum, 53

radiation pattern, 27
random signal, 59
random surface, 59
random turbulent velocity, 125
random uncorrelated contribution, 59
range marker, 77
Rayleigh distribution, 50
reciprocal transmission media, 37
rectangular distribution, 101
reduced mass, 56
reflectance spectrum, 61, 62, 64
reflecting surface, 21
refractive index integral, 49
residue intensity, 125
resolution cell, 37
resonance phenomena, 55
resonant conditions, 98
restoring force, 56, 150
Richardson number, 113
rigid object, 12, 74
rigid ship, 142
rigid target, 79, 140
rigidity factor, 54, 90
rigidity, 50, 75, 83, 84, 89
rigidly moving system, 52
robot vision, 4
roll, 135
rough boundary, 110

rough scattering, 101
rough surface scattering, 80
rough surface, 49, 69

SAR antenna, 34
SAR integration distance, 30
SAR integration range, 30
SAR, 6, 29, 83, 117
scalar properties of internal waves, 156
scale selective motion patterns, 2, 84, 89
scan-time limited bandwidth, 62
scattered field, 26
scattering cross-section, 92
scattering centres, 16, 19
scattering cross-section, 11, 16
scattering elements, 17
scattering matrix, 37, 55
scattering volume, 45
scintillation spectrum, 47
Scotch pine, 85
sea clutter, 92, 131, 140
sea state, 137
sea surface scattering, 92
SEASAT, 144
selective detection, 61
self-synchronizing system, 158
semi-rigid body, 139, 141
separation frequency, 76
shadowing effects, 11, 113
shallow thermocline, 96, 145, 154
shallow water, 146
Shannon's sampling theory, 89
shape of target, 135
shear forces, 111
shear mechanisms, 108
ship 'surfing', 136
ship targets, 131
ship wakes, 144
side-looking antenna, 25
sidebands, 20, 88
signal enrichment, 59
signature domain, 54
signature of the ocean surface, 120
solid scattering objects, 68
source area, 107
space-time spectrum, 92
space/time coherence, 51, 54, 83, 89, 117, 131
space/time mutual coherence, 143
spaced antennas, 88
spaced frequencies, 88
span of the scattering matrix, 38
spatial correlation distance, 59
spatial correlation, 15, 16, 24, 47, 72

spatial delay function, 14
spatial interferometry, 25
spatial radar signature, 51
spatial stationarity, 14
spectra of wave length, 124
spectral intensity, 86
specular reflection, 69
spread spectrum, 23
spring constant, 56
stationary, 14
statistical signal retrieval methods, 59
statistically orthogonal, 6, 69
stern waves, 146, 150
stimulating frequency, 56
stochastic functions, 13
stochastically variables, 15
stress characterization, 84
stress factors, 84
structure of the ionosphere, 111
sub-surface objects, 150
surface acoustic wave technology, 158
surface chemistry, 67
surface disturbances, 149
surface gravity waves, 96, 149, 153
surface roughness, 67
surface spectroscopy, 55, 57
surface waves, 97
surface wind, 59
symmetry characteristics, 71
symmetry of the target, 36
synchronous tuning, 12
synthetic aperture antenna, 25
synthetic aperture radar, 115
synthetic aperture, 24

tailored illumination, 4
target signature, 82
target/clutter enhancement, 141
temporal correlation properties, 45
temporal distribution, 70
temporal Fourier transform, 14
temporal integration, 45
temporal properties, 44, 48, 132
temporal spectra, 99
temporal variation, 70
terrain surface, 22
terrestrial background, 5, 6, 69
textural signature, 88
texture analysis, 86
texture, 84
thermocline, 150
time/space resolution cell, 72
topography of the lunar surface, 112

transfer function, 60
translatory motion 12, 51, 75, 135
transmission medium, 69
transverse distribution, 24, 70
transverse resolution, 30
transverse shape, 25
truncation influence, 101
truncation, 104
turbulence effects, 113
turbulence scale, 59
turbulence theory, 108
turbulence, 92
turbulent intensity, 111
turbulent irregularities, 108, 109
turbulent irregularity structure, 95
turbulent wake, 146
turbulent wave, 150
two layered ocean, 150
two-dimensional broadside array, 68
two-dimensional properties of internal waves, 156
two-lobed angular response, 109
two-peak spatial spectrum, 127

uncorrelated noise, 61
unidirectional waves, 101
up-convertors, 73

V-shaped shipwaves, 144
V-shaped surface gravity waves, 149

vegetation species, 84
vegetation, 21
velocity distribution, 46, 70
velocity shear, 113
vibration energy levels, 56
vibration frequency, 89, 91
vibrational quantum number, 56
video recording, 86, 89

wake pattern, 148, 149
wave direction, 122
wave height, 92, 93, 111
wave length resolution, 104
wave-height spectra, 8, 93, 95, 118
wave-number matching, 8
wave-tunnel measurements, 130
waveforms, 63
waveguide filter, 10
wavelength dependence, 20
wavelength domain, 62
waverider, 123
waves and turbulence, 93
white capping, 111, 128
wind gust pattern, 107
wind turbulence, 108
wind-tunnel experiments, 129

yaw, 135

zooming SAR, 29